U0621130

国家社科基金艺术学重大项目『中华民族服饰文化研究』

敦煌服饰文化图典

初唐卷

刘元风　赵声良　主编

Research on Chinese National Costume Culture Funded by Major Projects of National Social Science Foundation for Arts

The Illustration of Dunhuang Costume Culture

The Early Tang Dynasty

Editor-in-Chief
Liu Yuanfeng
Zhao Shengliang

中国纺织出版社有限公司

内 容 提 要

丛书将选择敦煌历代壁画（尊像画、故事画、经变画、史迹画、供养人像等）和彩塑中的典型人物形象，包括佛国世界中的佛陀、菩萨、弟子、天王、飞天、伎乐人，以及世俗世界中的国王、王后、贵族、平民、军队等，对其反映的服饰造型和图案进行整理绘制，并对其文化内涵进行理论研究。

书中每一单元的内容包括敦煌典型洞窟的壁画或彩塑原版图片、根据此图像整理绘制的服饰效果图和重点图案细节图，以及重要图像的服饰复原图。书中还收录了与主题相关的学术论文，并对对应图像的历史背景、服饰特征、艺术风格等方面进行了深入研究和概括说明。

本书适合院校师生、科研人员、设计师和敦煌文化艺术爱好者学习借鉴，同时具有一定的典藏价值。

图书在版编目（CIP）数据

敦煌服饰文化图典. 初唐卷 / 刘元风，赵声良主编. -- 北京：中国纺织出版社有限公司，2022.1
ISBN 978-7-5180-8793-8

Ⅰ. ①敦… Ⅱ. ①刘… ②赵… Ⅲ. ①敦煌学—服饰文化—中国—唐代—图集 Ⅳ. ① TS941.12-64

中国版本图书馆 CIP 数据核字（2021）第 162636 号

DUNHUANG FUSHI WENHUA TUDIAN CHUTANGJUAN

责任编辑：孙成成　　责任校对：王花妮　　责任印制：王艳丽

中国纺织出版社有限公司出版发行
地址：北京市朝阳区百子湾东里A407号楼　邮政编码：100124
销售电话：010—67004422　传真：010—87155801
http://www.c-textilep.com
中国纺织出版社天猫旗舰店
官方微博http://weibo.com/2119887771
北京雅昌艺术印刷有限公司印刷　各地新华书店经销
2022年1月第1版第1次印刷
开本：889 × 1194　1/8　印张：41.5
字数：392千字　定价：880.00元

凡购本书，如有缺页、倒页、脱页，由本社图书营销中心调换

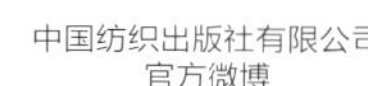

中国纺织出版社有限公司
官方微博

中国纺织出版社有限公司
官方微信

序一

象征着中西文化交流和友好往来的“丝绸之路”，是古代人民用巨大的智慧创造出来的一条光彩夺目的美丽缎带，而敦煌莫高窟则是镶嵌在这条缎带上的一颗闪闪发光的明珠。在延续一千多年的敦煌石窟艺术中，各个历史时期的彩塑、壁画人物的服装和服饰图案非常丰富精彩，是我们取之不尽、用之不竭的艺术源泉。

我非常有幸，自小跟随父亲常书鸿在敦煌学习壁画临摹，后来又因梁思成、林徽因两位先生走上了工艺美术的人生道路。父亲曾经提醒我：“不要忘记你是敦煌人，……也是应该把敦煌的东西渗透一下的时候了!”多年来，我一直盼望对包括服饰图案在内的敦煌图案进行全面系统的研究。1959年，我和李绵璐、黄能馥两位老师曾去敦煌石窟专门收集整理服饰图案，这些原稿在20世纪80年代首先由香港万里书店出版为《敦煌历代服饰图案》一书。书中的图案部位示意图，是当时刚留校不久的刘元风老师和他的同学赵茂生老师帮忙绘制的。20世纪90年代，我又带领我的研究生团队对敦煌图案进行了分类整理，相继出版了《中国敦煌历代装饰图案》和《中国敦煌历代装饰图案（续编）》，完成了我的部分心愿。

2018年6月，敦煌服饰文化研究暨创新设计中心在北京服装学院正式挂牌成立。中心有两个主要任务：一是把敦煌服饰文化艺术作为专题进行深入化和系统性研究，二是根据研究成果和体会进行创新性设计运用，涵盖了继承与发展的永恒主题。围绕着这两个主要任务，中心举办并展开了学术论坛、科研项目、人才培养、设计展演等一系列具有学术高度且富有社会影响力的活动，获得了院校、科研院所及行业的广泛好评。

此次由北京服装学院和敦煌研究院两家单位合作推出的“敦煌服饰文化图典”系列丛书，集中了各自在敦煌学和服饰文化研究方面的专长和优势，在理论和实践两个方面进行继承和创新。一方面，以历史时代的划分进行敦煌石窟历史背景和服饰文化的理论阐述，另一方面，基于理论研究展开敦煌石窟壁画和彩塑服饰的艺术实践。丛书不仅以整理临摹的方式将壁画和彩塑的整体服饰效果和局部图案进行整理绘制，同时也在尊重服饰历史和工艺规律的基础上，将壁画和彩塑中的人物服饰形象进行艺术再现，使之更加直观生动和贴近时代。

敦煌历代服饰文化的研究和继承工作，还仅仅是一个开始，绝不是结束。我希望更多的年轻学者、设计师、艺术家、学子们，能够继续前行，继续努力，做出更多的成绩，为弘扬我们引以为豪的敦煌艺术贡献自己的力量!

原中央工艺美术学院（现清华大学美术学院）院长

2020年9月

Preface 1

The Silk Road, which symbolizes the cultural exchange and friendship between China and the West, is a dazzling and beautiful ribbon created by the ancient people with their great wisdom. Mogao Grottoes is a shining pearl inlaid in this ribbon. It has lasted more than one thousand years, the costumes and patterns on costumes of painted sculptures and mural figures in various historical periods are very rich and wonderful, which is our inexhaustible artistic source.

I was very lucky to learn mural copying at Dunhuang with my father Chang Shuhong when I was a child. Later, I took the road of arts and crafts because of Mr. Liang Sicheng and Mrs. Lin Huiyin. My father once reminded me: "Don't forget that you are from Dunhuang, you should study Dunhuang deeper..." For many years, I have been looking forward to a comprehensive and systematic study of Dunhuang patterns, including clothing patterns. In 1959, two scholars, Mr. Li Mianlu and Mr. Huang Nengfu and I went to Dunhuang Grottoes to collect and arrange clothing patterns. In the 1980s, these manuscripts were first published by the Hong Kong Wanli Bookstore as the book *Costume Patterns from Dunhuang Frescoes*. The sketch illustration in the book was drawn by Mr. Liu Yuanfeng and his classmate Mr. Zhao Maosheng. In the 1990s, I led my post graduate team to sort out Dunhuang patterns, and successively published the books *Decorative Designs from China Dunhuang Murals* and *Decorative Designs from China Dunhuang Murals*(*continued*)(*sequel*), which fulfilled part of my wish.

In June 2018, Dunhuang Costume Culture Research and Innovation Design Center was officially established in Beijing Institute of Fashion Technology. The center has two main tasks: one is to take Dunhuang costume culture and art as a special topic for in-depth and systematic research; the other is to carry out innovative design and application according to the research results and experience, covering the eternal theme of inheritance and development. Around these two main tasks, the center has held and launched a series of activities with academic height and social influence, such as academic forum, scientific research project, talent training, design exhibition and so on, which has been widely praised by colleges, research institutes and the industry.

The Illustration of Dunhuang Costume Culture series jointly launched by Beijing Institute of Fashion Technology and Dunhuang Research Academy focuses on their respective expertise and advantages in Dunhuang study and clothing culture research, inherits and innovates in theory and practice. On the one hand, the historical background and costume culture of Dunhuang Grottoes are expounded theoretically according to the division of historical times. On the other hand, the art practice of Dunhuang Grottoes murals and painted sculptures costumes are carried out based on theoretical research. The series not only arrange and draw the overall clothing effect and patterns of murals and painted sculptures in the way of sorting and copying, but also on the basis of respecting the clothing history and technological rules. The series also make artistic reproduction of the characters' clothing images in the murals and painted sculptures, so as to make them more intuitive, dynamic and close to the times.

This research and inheritance work of Dunhuang costume culture is just a beginning, by no means the end. I hope that more young scholars, designers, artists and students could continue moving forward, working hard and making more achievements to contribute the promotion of Dunhuang art which we are proud of !

Chang Shana

Former president of Central Academy of Arts and Crafts

(now Academy of Fine Arts, Tsinghua University)

September 2020

序二

一、敦煌服饰文化的历史棖脉和艺术风格

敦煌石窟艺术博大精深，涵盖并综合了四至十四世纪、自十六国至元代十个历史时期的壁画、彩塑和建筑精品，是佛教艺术与中华民族传统文化相结合的集中体现，是世界闻名的文化遗产。

敦煌佛教艺术源于民族文化与世俗生活，体现着民间生态与佛教故事、西域艺术之间的交互融合。它以佛教故事为载体，描绘着世俗百态和人间万象，反映着古代社会生活和生产中衣、食、住、行的情景。敦煌石窟壁画和彩塑的人物造像生动自然、形态优美，服饰精致、图案华丽。同时，敦煌服饰因不同时期的文化形态和生活样貌的变迁，在其造型、结构、材质、配饰、色彩、纹样和装饰风格方面也各不相同。常书鸿先生曾在《敦煌历代服饰图案》一书的序言中写道："敦煌艺术，不仅反映了外来文化影响和隋唐盛世的佛教美术以及当时的社会生活，而且记录了中国历代的装饰图案、色彩运用和工艺技术。从敦煌壁画和彩塑上临摹下来的丰富多彩的纹样图案，实际上就是中国历代服饰和织造、印染工艺的重要历史资料。"因此可以说，敦煌石窟艺术的发展史同时也是一部宗教性与民族性高度融合的社会文化生态史。

十六国至北周时期的敦煌石窟壁画和彩塑多以土红为底，饱满热烈。由于颜料变色的缘故，人物造型大多显得古朴粗犷，肌肤与面部着色受到西域艺术凹凸画法的影响，给人以很强的厚重感和立体感。佛教人物的服饰更多地受到印度和波斯的影响，而世俗人物的服饰多带有魏晋时期褒衣博带、曲裾飞髾的飘逸之感，袴褶的流行为隋唐服饰多元化奠定基础，服饰图案以忍冬纹和几何纹为主。忍冬纹多为三瓣叶或四瓣叶构成的二方连续纹样，在敦煌早期洞窟的藻井和壁画的边饰中作为主体装饰，在此时人物服装的领子、门襟、袖口等部位上出现了相应风格的装饰。以回形、菱形、山形为主要结构的几何纹反映了当时的提花织造技术，形成了这一时期石窟造像服饰艺术的重要风格。

隋唐时期是敦煌石窟艺术发展的鼎盛时期，也是佛教文化空前活跃并与当时的生活、艺术结合最为紧密的时期。隋代壁画和彩塑人物造型颀长，服饰风格方面承上启下，注重面料图案和质感的表现，显示了由朴实简约向奢华盛装过渡的特点，以及当时染织工艺的交流和提高。此时彩塑菩萨裙装上反复出现的联珠纹，即在珠状圆环或菱形骨架中装饰狩猎纹、翼马纹、凤鸟纹、团花纹的四方连续或二方连续纹样，便是在受到波斯萨珊王朝装饰风格影响基础上进行本土化创造的产物。唐代壁画和彩塑人物造像愈加生动写实，造型日趋丰腴，服饰形态多样，图案描绘细致，不仅再现了天子、王侯、官吏及百姓等各社会阶层的服饰礼制，以及各国、各民族的服饰，而且晚唐出现的供养人像服饰，突出反映了当时最奢华的流行时尚，同时也折射出此时社会经济繁盛、中西文化交融的特点。例如，盛唐第130窟都督夫人太原王氏供养人像，描绘了盛唐时期贵族妇女体态丰盈，着襦裙、半臂、披帔帛，服饰上装饰着清新妩丽的折枝花纹的生动景象。随侍的侍女着圆领袍服，束革带，反映着当时女着男装的流行现象。此时的服饰图案更加丰富逼真，如卷草纹、宝相花纹等融合了希腊艺术中的莨苕叶、佛教文化中的莲花、丝绸之路上的石榴等多种装饰基因，成为当时各国文化交流的历史见证。而通过画师画笔表现出来的织锦、印花、绞缬等多姿多彩的染织工艺，更说明唐代丝绸技艺在世界历史上的重要地位。

敦煌石窟的整体发展在五代之后步入后期，艺术的创造力和感染力逐渐减少，许多壁画和彩塑造像出现了程式化的现象，但是对于服饰文化来说，这段时间却是异彩纷呈、百花争艳。因为此时供养人像在壁画中所占比重大幅度增加，且人物身份地位丰功显赫，成为画师们重点描绘的对象，其中女供养人的服饰和妆容十分精致，如五代第98窟曹氏家族女供养人像，由钿钗、大袖衫、长裙、帔帛所组成的花

钗礼服，真实反映了贵族妇女沿袭唐代且更加繁缛的奢华服饰。由于多民族聚居和交往的历史背景，此时壁画中还出现了于阗国王和皇后、回鹘王和王妃、回鹘公主等具有民族风格的服饰，而且对西夏党项族和元代蒙古族的服饰也有表现。这些珍贵的服饰图像均具有珍贵的历史价值，充分反映了这一时期民族融合的多元性。

敦煌保存的壁画和彩塑中所呈现出来的服饰文化风貌，无疑是中华服饰历史的重要组成部分，也是中国中世纪服饰文化的艺术宝库，反映了中国古代人民杰出的艺术才能和生活智慧。这些资料构成了一座中华民族服饰文化和装饰艺术的博物馆，为艺术设计和理论研究的工作者们提供了取之不尽、用之不竭的学习研究资源。

二、敦煌服饰文化的学习传承和设计创新

我于1977年就读于中央工艺美术学院（现清华大学美术学院）染织设计专业，当时的班主任是常沙娜老师。常沙娜老师是被誉为“敦煌守护神”的敦煌研究院第一任院长常书鸿之女，少年时曾跟随父母在敦煌莫高窟学习临摹壁画艺术。大学期间，常沙娜老师请来常书鸿先生为我们做过两次学术讲座。听着常书鸿先生用特有的杭州口音讲述敦煌石窟艺术，特别是敦煌图案在建筑、服饰等各个方面的丰富体现，引起了我对敦煌艺术的兴趣，也奠定了我对敦煌历代服饰艺术的初步印象。更加幸运的是，在我们上大二时，常沙娜老师为我们讲授敦煌服饰图案课程，那年的暑假又带领我们去敦煌莫高窟实地考察。对于我们这群学习染织专业的学生来说，一进到石窟中亲眼看到千百年前遗留下来的璀璨壁画和彩塑，那些鲜活生动的人物形象、多姿多彩的服饰造型、华丽精致的面料体现……都让我们如同置身时间的万花筒，真切领略到蕴藏在线条和色彩中的敦煌服饰文化的无穷魅力。

除了赴敦煌石窟实地领略敦煌艺术的机缘外，另外一段工作经历也促成了我对敦煌服饰文化的认知和热情。1980年，常沙娜老师整理的三百余幅敦煌历代服饰图案画稿准备出版。这批画稿是20世纪50年代，由常沙娜老师、李绵璐老师、黄能馥老师赴敦煌莫高窟考察整理绘制而成的，十分宝贵，但由于各种原因，直到20世纪80年代初才有机会出版。为了更好地展现敦煌历代服饰图案的全貌，常沙娜老师要求我找出画稿所对应的壁画或彩塑，整理绘制成人物线稿图，并在图上标识出图案的装饰部位，令读者一目了然。于是，我花了近三个月的时间，潜心研究、整理、绘制这批人物线稿图，最后用针管笔细细勾勒成定稿。这部书最终在1986年由香港万里书店有限公司和轻工业出版社正式出版，后来在2001年和2018年又分别由中国轻工业出版社和清华大学出版社再版，至今畅销不衰，在国内外服饰历史研究、装饰设计、服装设计、影视剧舞台美术等领域发挥了极为重要的作用。囿于时间仓促，现在看来由我负责绘制的这批人物线稿图，在局部造型和服饰穿插等方面还有一些不尽完美的地方，但是这项工作给予我深入学习敦煌服饰文化的难得机会，为我指明了学习和运用传统服饰文化的新途径，也为我在后来的服装教育和设计工作中进行传承和创新的实践型研究奠定了重要基础。

应社会发展需求，1980年染织设计系开设了服装设计专业。1982年我毕业后留校任教，随着1984年成立服装设计系，根据学校安排和个人意愿，在服装设计系担任授课教师。一方面，我在授课中向学生们积极介绍敦煌服饰艺术的历史渊源和风格特征，传播和发扬优秀的中华服饰文化；另一方面，通过设计实践不断磨炼和提升个人的专业素养，我对敦煌服饰文化的认知也不断加深。1992年香港时装文化节上，我的“敦煌系列”“陶瓷系列”“剪纸系列”等主题性时装作品参展；1996年敦煌山庄设计项目中，我参与设计的工作服方案得到采纳；2001年清华大学九十年校庆“艺术与科学”国际研讨会“时尚·交融”服装展演中，我运用敦煌元素设计的系列服装受到国内外嘉宾的好评……2002年，我调入北京服装学院工作，继续在新的工作岗位利用和发挥科研平台及时尚传播的影响力，在2013年与敦煌研究院合作组织举办了“垂衣裳——敦煌服饰艺术展”，通过展览、服饰秀、论坛等一系列学术活动，集中展示了

跨越千年的敦煌服饰艺术精华，促使优秀传统文化走进高校、走进现代、走进生活，迸发出强盛的生命力，受到社会各界的一致赞赏。

2017年，敦煌服饰文化的发展迎来新的契机。12月7日，受中华人民共和国文化和旅游部邀请，作为中英高级别人文交流机制第五次会议的重要内容之一，北京服装学院与敦煌研究院、英国王储传统艺术学院、敦煌文化弘扬基金会共同签署“敦煌服饰文化研究暨创新设计中心”战略合作框架协议，我国及英国多位领导人共同见证了协议签字仪式。2018年6月6日和2020年1月6日，“敦煌服饰文化研究暨创新设计中心”的挂牌仪式分别在北京服装学院和敦煌研究院正式举行。

敦煌服饰文化研究暨创新设计中心成立的主旨是综合四方在敦煌文化艺术研究教育、文化传承、创新设计、社会传播、大众推广、国际合作与交流等方面的优势，汇集敦煌与丝绸之路文化研究、服饰文化研究、艺术设计、传统技艺、传播推广等各方面的人才与资源，深入开展敦煌服饰文化研究、当代创新设计和研发、研究和创新设计成果推广，搭建敦煌服饰文化艺术保护研究、文化传承、创新设计、人才培养与交流、社会传播、产业化转化等的国际化学术平台。自中心成立以来，中心研究团队成员在理论研究、人才培养、文化推广等各方面取得了一系列成果，包括承担2019国家艺术基金“敦煌服饰创新设计人才培养”项目、2019国家社科基金艺术学项目“敦煌历代服饰文化研究”、第三届和第四届丝绸之路（敦煌）国际文博会“绝色敦煌之夜”展演项目，举办敦煌服饰文化论坛和系列学术讲座等。“敦煌服饰文化图典”系列就是中心在多年潜心研究的基础上，发挥专业优势，面向院校师生、科研人员、设计师和喜爱敦煌艺术的读者所推出的重要学术成果。

三、《敦煌服饰文化图典》的编纂体例和研究工作

本书采取图文并茂的形式，依据敦煌石窟所辖十个历史时期以及对应的服饰内容体量进行体例划分。图典不仅涵盖敦煌石窟营造历时千年的伟大历程，着重展现敦煌历代服饰文化的面貌，而且根据敦煌石窟建造的时代特点和发展规律，突出唐代等石窟数量较多、图像资料较为丰富的时期的服饰面貌，重点呈现最具有代表性的时代和洞窟中的珍贵资料，凸显敦煌服饰文化的独特价值。

在人物形象的选取方面，本书聚焦敦煌历代壁画（尊像画、故事画、经变画、史迹画、供养人像等）和彩塑中的典型人物形象，包括佛国世界中的佛陀、菩萨、弟子、天王、飞天、伎乐人等，以及世俗世界中的国王、王后、贵族、平民、侍从等，对其反映的服饰造型、图案、色彩和文化内涵进行图像绘制和理论阐述。由于壁画和彩塑的时代特征鲜明，造像风格各异，所以书中对历代人物身份的侧重也不同。例如，敦煌早中期石窟中服饰绘制着墨较多的是佛陀、菩萨、弟子等人物形象，而晚唐之后石窟壁画中出现了大量供养人像，其服饰大多具有历史真实来源且描绘精致，民族风格突出，因此作为相应卷宗的主体内容进行表现。

书中每一单元的图版部分包括由敦煌研究院提供的敦煌壁画或彩塑原版高清图片，以及根据此图像整理绘制的人物服饰效果图和重点图案细节图。文字部分为关于此图像的理论研究，主要从敦煌石窟的历史背景、服饰特征、绘画风格三个主要方面进行分析，而且对照同时期的文献资料和绘画、雕塑、服饰纺织品等图像和实物资料进行类比研究和严谨考证，做到言之有据、言之有物。

自2018年起，北京服装学院和敦煌研究院的研究团队开始着手进行本书的编纂工作。按照本书的总体规划和具体实施计划，首先依据研究人员的专业背景和研究方向进行任务分工，主要分为绘图组、理论组和艺术再现组这三个小组。

绘图组由我总体统筹，负责书中的图版绘制工作，根据绘图内容和专业侧重，又细分为人物服饰效果图组和服饰图案组。人物服饰效果图组成员的学术背景多为服装设计和理论方向，具有扎实的造型基础和色彩把握能力，能够较好地表现人物和服饰的整体关系。服饰图案组成员的学术背景多为染织服

装设计和理论方向，具有较好的图案构成和装饰色彩基础，能够充分表现服饰图案的整体造型、构图和色彩。

理论组由敦煌研究院赵声良院长总体把关，负责书中的论文和图版说明写作工作。成员的学术背景多为佛教美术史和服装史方向，能够从敦煌石窟的时代背景和脉络渊源出发，阐明服饰文化整体趋向和服饰艺术个案特色的学术价值。

艺术再现组由北京服装学院楚艳教授总体负责，成员多具备丰富的服装历史和设计工艺等专业积累，能够在理论研究和实践经验的基础上，通过服装结构解析、纹样整理、面料织造、色彩染制等工艺技术和妆容造型等因素作用，探索从壁画平面绘制到现实立体再现的接续和跨越，对敦煌服饰艺术进行分析、整理、把握和创造。

团队的研究方法是以敦煌石窟实地考察为基础，结合文献资料查证，以珍贵的一手资料和艺术感受为前提的艺术实践型研究。本书中图版所占分量很大，而且图片的绘制风格和最终效果也是影响图书品质的第一要素。所以在成书过程中，团队对于图版的绘制风格曾经过反复论证和考量。因为这些图版的绘制不同于完全依照敦煌壁画现状的复制，也不同于敦煌壁画原状的复原，而是在忠于敦煌壁画造型、色彩和构图的基础上，运用服装语言和图案学的组织构成原理，同时加入绘制者对图像的新的理解和新的诠释，并将残缺部分进行合理补充，将褪色部分进行科学完善，最终再现并赋予敦煌服饰文化以新的面貌。

在明确了研究方法和绘图风格后，团队中的绘图组根据任务分工开始进行绘制图稿，然后多次深入敦煌石窟实地考察和图像比对进而完善和修正，在绘制过程中，对某些局部造型、结构等进行适度的调整，直到最后完成图稿。在此期间，理论组通过实地调研和资料收集，用简明扼要的文字对图像所在洞窟、壁画的艺术风格和服饰特色进行整理撰写，同时和绘图组相互沟通和反复校对，在图版文字中加入绘图人员的经验和感受，最终完成写作。通过大家的不懈努力，在2020年完成了阶段性的研究，推出此部《敦煌服饰文化图典　初唐卷》，即敦煌初唐时期服饰文化的研究成果。

在本书的编纂过程中，我和团队人员深切感受到敦煌石窟艺术创造者的高超智慧和敦煌服饰文化的博大精深，深刻认识到这项研究工作不是一蹴而就的，而是需要时间积累和沉淀的。团队成员时常为了人物的优美造型反复修改，为了服饰色彩的典雅沉着进行多次调试，为了说明文字的准确简练琢磨炼句。事实证明，只有经过长时间的反复学习和揣摩，有了亲身体验和历练，才会有新的发现和新的启示，才能吸收好和利用好这座人类艺术的宝库。

四、敦煌服饰文化的研究意义和发展方向

1900年敦煌藏经洞被发现，数量众多、价值连城的石室宝藏震惊世界。1944年成立敦煌研究院以来，经过几代学人的努力，敦煌学已经取得了丰厚的研究成果，发展成为名副其实的国际性显学。但就目前敦煌学的研究现状来看，多集中于历史学、考古学、世界史、民族学、语言学等学科，关于敦煌服饰领域的专门研究还较为稀缺。因此，针对目前敦煌学研究中的薄弱环节，专门对敦煌历代服饰文化进行梳理和研究，无疑是对敦煌学研究的重要完善。此外，从中国古代服装史研究的角度来看，中国历代舆服志所载多为中原王朝官服和礼服制度，对于多朝代、民族、宗教背景下的丝绸之路沿线所反映的服饰历史和文化内涵研究还有所欠缺，关于历代各民族世俗百姓服饰文化的研究还十分匮乏。因此，敦煌服饰文化研究也是中国古代服饰文化研究领域中不可或缺的重要部分。

不同时代的学者和设计师对敦煌艺术的理解和感悟是有所不同的，当代学者和设计师对敦煌艺术的认知理应加入新时代的文化理念和艺术思潮以及社会风尚。除了对敦煌服饰文化进行潜心学习、研究和传承外，在新时代的背景下，社会经济的不断发展和民众对传统文化的喜爱和需求无疑会对服装历史研究学者和设计师们不断地提出新的课题。如何使古老的敦煌服饰艺术重归现代设计和大众生活，谱写出更加有活力的未来呢？根据前辈学者的研究和体验可以知道，运用现代设计的眼光和手法，汲取敦煌服饰艺术的深厚滋养，赋予装饰元素以时代感和创新性，古为今用，能够在现代设计舞台上充分展现敦煌的文化创新和时代魅力，有效推动传统文化的弘扬。例如，常沙娜老师在20世纪50年代创作出亚洲及太平洋区域和平会议礼品“和平鸽”丝巾，之后又在参与首都十大建筑的过程中大胆融入敦煌元素，在众多设计方案中脱颖而出。半个多世纪以来，这些敦煌元素设计经历了时间的考验，因其华美庄重、大气磅礴的风格受到国内外领导和大众的普遍肯定和赞赏，体现了以敦煌艺术为代表的中国优秀传统文化在现代设计语境下的永恒生命力。

“敦煌服饰文化图典”系列是在已有学习和研究的基础上，着重于服饰文化的视角重新审视，重新探究，重新提炼，从而得出一种新的服饰面貌和服饰形态，特别是对服饰造型和服饰结构的认知，对服饰图案及色彩的把握，都会有新的思考和新的诠释融入其中。让大众看到：久远的敦煌艺术正在走向当代社会生活；敦煌艺术既是古典文化的，更是现代时尚的，使敦煌艺术焕发出新时代的生机与活力。所以，通过本书的出版和传播，希望敦煌服饰文化能够被更多的艺术家、设计师、敦煌艺术的爱好者了解、喜爱和运用，引发更多关于传统文化与现代设计结合的思考。在构建中国传统服饰文化脉络和传承体系，以及指引现代文创产业和产品创新设计方面做一点实际工作，真正树立属于中国人的文化自信，构筑包容宏大的中国精神。

劉元風

北京服装学院　教授

敦煌服饰文化研究暨创新设计中心　主任

2021年1月

Preface 2

1. The historical roots and artistic styles of Dunhuang costume

Dunhuang Grottoes contain extensive materials, which cover murals, painted sculptures and fine architectures of ten historical periods from the 4th to 14th century AD, the Northern and Southern Dynasties to the Yuan Dynasty. They are a perfect embodiment of the combination of Buddhist art and traditional Chinese culture, a world-famous cultural heritage.

Dunhuang Buddhist art derived from both the national culture and the secular life, shows the interactions and fusions among folk culture, Buddhist story and Western Region's art. It uses Buddhist story as the carrier, depicting secular life and natural world, reflecting the ancient social life of clothing, food, housing and transportation. Dunhuang murals and painted sculptures are vivid in character, beautiful in form, elegant in clothes and delicate in pattern. Also they showed the process of Dunhuang costume evolving due to the change of culture and life in different times, manifested on the style, fashion, material, accessory, color, pattern and decoration motif. Mr. Chang Shuhong wrote in the preface of the book *Costume Patterns from Dunhuang Frescoes*: "Dunhuang art reflected not only the foreign culture influence, the Buddhist art during Sui and Tang Dynasty and social life, but also recorded the history of Chinese clothes, the decoration patterns, color combinations and production technology. In Dunhuang murals and on painted sculptures, the rich and colorful decorating patterns actually are the important historical records of the Chinese ancient costumes weaving and dyeing history." So it can be said that the history of Dunhuang grottoes art is also a social and cultural history ecology with the fusion of religious and national history.

During Northern and Southern Dynasties, Dunhuang grottoes murals and sculptures were usually painted in red as background, full of passion and energy. Due to oxidation, some colors especially the skin parts have changed, so the figures look wild and simple, the skins especially faces coloring technique influenced by Indian chiaroscuro painting style, give audiences a kind of heavy and perspective feeling. The Buddhist figures' costume are more influenced by Indian and Persian culture, and the secular characters' clothes more close to people's clothing style in Wei-Jin Period, which is quite loose, with long sleeves, and looks relax and elegant. The popular of the Kuzhe (袴褶) dress laid the foundation for the developing and diversity of Sui and Tang Dynasties costumes, and the clothing patterns were mainly used with honeysuckle and geometric lines. Honeysuckle pattern usually has three or four leaves, with two as a group connected with each other made up a kind of continuous ornamental motif as the main decoration elements on the ceilings and edges of murals in the early Dunhuang caves. At the mean time the patterns on clothes like collar, front open edges, cuffs also have similar decoration. Geometric patterns like hollow square which encircled with another hollow square, rhombuses, triangles reflects the jacquard weaving technology at that time, which constituted an important style of the grotto costume art in this period.

In Sui and Tang Dynasties Dunhuang grottoes art developed to its heydays, meanwhile Buddhism also was unprecedentedly active, infiltrated in life and art deeply. The figures in murals and painted sculptures in Sui Dynasty are usually tall and big, the clothes design in this period can be seen as a connecting link between the preceding and the following, started to pay more attention to patterns and manifestation of the texture, showing the transition process from simple and practical to complex and luxury style, it also indicates that the dyeing technique improved at that time. During this period, the clothes of painted sculpture Bodhisattvas repeatedly have Sassanian roundels motif, sometimes in a circle, sometimes in a rhombus shape, inside the geometries sometimes painted a hunting scene, or a winged flying horse, or a phoenix, or a flower pattern, two or four geometries as a group, continuous repeating to fill up the wanted space. This design is a kind of local recreation based on the influence of Sassanian decoration style. Figures in murals and sculptures of Tang Dynasty are more close to real people, vivid and more rounded. The clothes design are more diverse, with delicate patterns, and revealed people of that time like kings, officials, ordinary people in different social status have different clothing rules, also foreigners, minorities can be easily identified by their different characteristic costumes. The donor images from Late Tang Dynasty show the most luxury fashion at that time, and reflect the social background which was prosperous and diverse. For example in cave 130, there is a female donor portrait, painted in High Tang Dynasty, depicting vividly the aristocratic lady at that time with graceful, rich, well-shaped body, wearing a kind of shirt and dress, coated with half-sleeved garment, wrapped with silk shawl. The clothes are decorated with vivid scenes of bright and charming broken branch patterns. The attendants wore round necked robes and leather belts, which reflected the popular phenomenon of woman wearing man's clothes at that time. During this period the clothing patterns were more rich and realistic, such as scrolling grass patterns and Baoxiang（宝相）patterns, which integrated various decorative genes, such as Hyoscyamus leaves from Greek art, lotus flowers from Buddhist culture, pomegranates on the Silk Road and so on, these became the historical witness of cultural exchanges among countries at that time. Through the painters' brush pen, the colorful dyeing and weaving techniques of brocade, printing and tie-dye which are displayed clearly, reflecting the advanced silk production technique in Tang Dynasty.

Since the Five Dynasties, the overall development of Dunhuang Grottoes entered the later stage. The creativity and appeal of art gradually decreased. Many murals and painted sculptures became stylized. However, for the clothing culture, this period was colorful and competitive. Because at this period, the proportion of donors in the murals enlarged greatly, and the social position of the patrons were high and powerful, who became the new focus of the painters. Among them, the dress and make-up of female donors were very exquisite. For example, the female donors of the Cao family in cave 98 of the Five Dynasties, the full dress composed by jeweled hairpin, large sleeve shirt, long skirt and silk shawl, truly reflected the noble women clothing inherited from the Tang Dynasty and even more elaborate and luxury. Due to the historical background of multi-ethnic settlements and communications, the murals at that time also appeared the full-length portrait of Khotan king and queen, Uighur king and princess and other costumes with national style, and also showed the costumes of Tangut in Western Xia and Mongolian in Yuan Dynasty. These precious clothing images have precious historical value, which fully reflects the national integration and the diversity of this period.

Dunhuang murals and painted sculptures preserved the cultural features of clothing, which is undoubtedly an important part of the history of Chinese clothing, as well as the art treasure house of Chinese medieval clothing culture, reflecting the outstanding artistic talents and life wisdom of the ancient Chinese people. These materials constitute a museum of Chinese costume culture and decorative art, providing inexhaustible learning and research resources for art design and theoretical research workers.

2. Study, inheritance and innovation of Dunhuang costume

In 1977, I majored dyeing and weaving design in the Central Academy of Arts and Crafts（now the Academy of Fine Arts of Tsinghua University) . At that time, the head teacher of my class was Chang Shana. Chang Shana is the daughter of Chang Shuhong, the first president of Dunhuang Academy, known as the "Patron Saint of Dunhuang". When she was a teenager, she studied mural copying in Mogao Grottoes with her parents. When I was in the college, Ms. Chang Shana invited Mr. Chang Shuhong to give us two academic lectures. Listening to Mr. Chang Shuhong's narration of Dunhuang Grottoes art with his unique Hangzhou accent, especially the rich expression of Dunhuang patterns in architecture, clothing and other aspects, aroused my interest in Dunhuang art, and also established my initial impression of ancient Dunhuang costume art. Fortunately, when we were sophomores, Ms. Chang Shana taught us the course of Dunhuang costume patterns. During the summer vacation of that year, she led us to visit Dunhuang Mogao Grottoes. As the students majoring in dyeing and weaving, as soon as entering the grottoes, we could see through our own eyes the splendid murals and colored sculptures lasting thousands of years. The vivid figures, the colorful clothing shapes, and the gorgeous and exquisite fabrics manifestation, all make us feel like a kaleidoscope of time, and we can truly appreciate the infinite charm of Dunhuang clothing culture which lies in lines and colors.

Except the opportunity of visiting Dunhuang Grottoes to appreciate the Dunhuang art, another period of work experience also contributed to my understanding and enthusiasm for Dunhuang costume culture. In 1980, more than 300 drawings of Dunhuang costume patterns compiled by Chang Shana were ready for publication. These paintings were made by Chang Shana, Li Mianlu and Huang Nengfu in the 1950s. They were very precious, but not published until the early 1980s for various reasons. In order to show the whole picture of Dunhuang costume patterns better, Ms. Chang Shana asked me to sort out the corresponding murals or colored sculptures, organize and draw the outlines of the figures, and mark the decorative parts of the patterns on the drawings, so that the readers can understand it clearly. Therefore, I spent nearly three months studying, sorting out and doing the line drawings of these figures, with a needle pen I finally made the final draft. The book was finally published by Hongkong Wanli Bookstore Co. Ltd. and Light Industry Press in 1986. Later, it was reprinted by China Light Industry Press and Tsinghua University Press in 2001 and 2018 respectively. It has been selling well, and plays a very important role in the field of clothing history research, decoration design, fashion design, film and TV, drama stage art at home and abroad. Because of the shortage of time, it seems that there are still some imperfections in the partial modeling and clothing matching of the figures drew by me. However, this work has given me a rare opportunity to learn more about Dunhuang clothing culture, and has pointed out a new way for me to learn and apply the traditional clothing culture in my work. It also helps me to carry out the clothing education and design work in the future, laying an important foundation for practical research on inheritance and innovation.

In response to the needs of social development, the Department of Dyeing and Weaving Design set up the major Clothing Design in 1980. After graduating in 1982, I stayed in college to teach. With the establishment of the Department of Clothing Design in 1984, I worked as a teacher in the Department by the school's arrangement and my personal wishes. On the one hand, I actively introduced the historical origin and style characteristics of Dunhuang costume art to the students in the course of teaching, spread the Chinese costume culture; on the other hand, through the design practice, I constantly honed and improved my professional ability, and my knowledge of Dunhuang costume culture also deepened. In 1992, at the Hong Kong Fashion Culture Festival, my theme fashion works such as Dunhuang Series, Ceramic Series and Paper-Cut Series were exhibited. In the design project of Dunhuang Villa in 1996, my design for working suit was adopted. In 2001, at the Fashion and

Integration exhibition of Tsinghua University's 90th anniversary Art and Science international seminar, I used Dunhuang elements to design a series of clothes which were well accepted by guests at home and abroad. In 2002, I transferred to Beijing Institute of Fashion Technology and continued using research platform and fashion trend to propagate Dunhuang in my new job. In 2013, I organized a Hanging Dress — Dunhuang costume Exhibition in collaboration with the Dunhuang Research Academy. Through exhibitions, fashion shows, seminars and a series of academic activities, we showed the beauty of Dunhuang's dress art with a history of thousands of years. The good traditional culture has been introduced into universities, contemporary life. It has burst out strong vitality, and has been unanimously appreciated by all walks.

In 2017, the development of Dunhuang clothing culture ushered into a new era. On December 7, by the invitation of the Ministry of Culture, as one of the important contents of the fifth meeting of China-UK high level people to people exchange project, Beijing Institute of Fashion Technology, together with Dunhuang Research Academy, British Crown Prince Institute of Traditional Arts and Dunhuang Culture Promotion Foundation, signed a strategic cooperation framework agreement on Dunhuang Costume Culture Research and Innovation Design Center, vice premier Liu Yandong and British health minister Hunt witnessed the signing ceremony of the agreement. On June 6th, 2018 and January 6th, 2020, the unveiling ceremony of Dunhuang Costume Culture Research and Innovation Design Center was officially held both at Beijing Institute of Fashion Technology and Dunhuang Research Academy.

The main purpose of the establishment of Dunhuang Costume Culture Research and Innovation Design Center is to integrate all hands interested in Dunhuang culture from art research and education, cultural heritage, innovative design, social communication, mass promotion, international cooperation and exchange, and to collect the talents from Dunhuang and Silk Road culture research, clothing culture research, art design, traditional skills, communication and promotion, etc resources, for in-depth development of Dunhuang clothing culture research, contemporary innovative design, research and development, research and innovation design achievements promotion, to build Dunhuang clothing culture and art protection research, cultural heritage, innovative design, personnel training and exchange, social communication, industrial transformation and other international academic platform. Since the establishment of the center, members of the research team of the center have made a series of achievements in theoretical research, talent training, cultural promotion and other aspects, including undertaking the 2019 National Art Fund Dunhuang Clothing Innovation and Design Talent Training Project, 2019 National Social Science Fund Art Project Dunhuang Clothing Culture Research, holding the third and fourth Silk Road (Dunhuang) International Culture Research Project, the exhibition project of Wonder Dunhuang Night, and Dunhuang clothing culture forum and a series of academic lectures. This series The Illustration of Dunhuang Costume Culture is an important academic achievement of the center on the basis of many years' dedicated research, giving full play to the professional advantages, suit for college teachers and students, researchers, designers and readers who love Dunhuang art.

3. The compilation and research work of this book

This book is illustrated with pictures. It not only covers the great process of the construction of Dunhuang Grottoes for thousands of years, but also highlights the clothing features of the Tang Dynasty and other periods when there are many caves and rich image materials according to the characteristics and development rules of the times. It focuses on the most representative era and most precious materials in caves, highlighting the unique value of Dunhuang clothing culture.

In the selection of characters, this book focuses on the typical characters in Dunhuang murals (such as Buddha image paintings, story paintings, scripture paintings, historical event paintings, patrons' portraits, etc.) and painted sculptures in Dunhuang, including Buddha, Bodhisattva, disciples, heavenly kings, flying Apsaras, musicians, and kings, queens, nobles, civilians and attendants in the secular world. We elaborate the style, pattern, color and cultural connotation by image drawings and introductions. Because of the distinctive characteristic of the times and the different styles of the sculptures, the emphasis on the identity of the characters in different dynasties are also different. For example, in the early and middle period of Dunhuang Grottoes, the images of Buddha, Bodhisattva and disciples were mostly painted and the craftsman gave much attention to their costumes. However, after the Late Tang Dynasty, a large number of patrons appeared in the murals, most of their costumes have historical origins and are well presented, and their national styles are outstanding. Therefore, they are expressed as the main content of the corresponding volumes.

The illustration part of each unit in the book were drawn according to the original high-definition pictures of Dunhuang murals or colored sculptures provided by Dunhuang Research Academy, as well as the design sketch and key pattern details of figures' clothes. The text part is about the theoretical study of this image, mainly analysing three aspects: the historical background of Dunhuang Grottoes, cothes characteristics, painting styles, and comparing with the contemporary literature and painting, sculpture, cothes textile for comparative research and check, to make this book professional and reliable.

Since 2018, the research team from Beijing Institute of Fashion Technology and Dunhuang Research Academy began to compile this book. According to the overall plan and specific implementation plan of this book, firstly, according to the professionals' backgrounds and research directions of the researchers, the task is divided into three groups: drawing group, theory group and art recreation group.

The drawing group is led by me who is responsible for the illustration drawing work in the book. According to the drawing contents and professional requirements, it was subdivided into the effect drawing group of character's clothing and the clothing pattern group. The academic background of the group members mostly are clothing design and theoretical direction. They have solid modeling skill foundation and the ability of color managing, who could express the overall relationship between characters and clothing clearly. Most members of the clothing pattern group have academic background of dyeing and weaving clothing design and theoretical direction. They have good pattern composition skill and decorative color study foundation, could fully depict the clothing pattern's overall modeling, composition and color.

Zhao Shengliang, Chief director of Dunhuang Research Academy, is responsible for writing articles and interpretations in the book. Most of the members' academic backgrounds are Buddhist art history and clothing history. They can clarify the overall trend of costume culture and the academic value of clothing art from the history background and origins of Dunhuang Grottoes.

The art recreation group is led by professor Chu Yan from Beijing Institute of Fashion Technology. Most of her members have rich professional working experience in clothing history and design technology. Based on theoretical research and practical experience, the team can explore from the plane drawing murals to the three-dimensional reality through the analysis of clothing structure, pattern arrangement, fabric weaving, color dyeing and other process technology, to analyze, arrange, grasp and create the new Dunhuang clothing art.

The team's research method is based on the field investigation of Dunhuang Grottoes, combined with scripture verification, and rely on the precious first-hand materials and artistic feelings as the premise to do the practical research. In this book, the picture accounts for a lot of weight, and the drawing style and final effect of picture are also the first factor affecting the quality of the book. Therefore, in the process of compiling, the team has repeatedly demonstrated and considered the drawing style of the pictures. Because the drawing of these pictures is different from copying exactly according to the status quo of Dunhuang murals, and also different from the restoration of the original Dunhuang murals, but being loyal to the shape, color and composition of Dunhuang murals as premise, the principles of clothing language and patternology are applied, at the same time, the new understanding and interpretation of the images by the painters are added, and the incomplete parts are reasonably supplemented, the fading part was scientifically recovered, and finally recreating and endowing Dunhuang clothing culture with a new look.

After clarifying the research methods and drawing styles, the drawing group of the team started drawing according to the requests, and then visited Dunhuang Grottoes many times to improve and revise the images. In the process of drawing, some shapes and structures were moderately adjusted until the final draft was completed. During this period, through field research and data collection, the theory group sorted out and wrote the artistic styles and clothing characteristics of the caves and murals in simple and concise words. At the same time, they communicated with the drawing group and proofread them repeatedly, added the experience and feelings of the painters into the text, and finally completed the writing. Through our unremitting efforts, we have completed the phased research in 2020, and launched this book *The Illustration of Dunhuang Costume Culture — the Early Tang Dynasty*, which is the research results of Dunhuang clothing culture of the early Tang Dynasty.

In the process of compiling this book, my team and I deeply realized the wisdom of the creators of Dunhuang Grottoes and the breadth and profundity of Dunhuang clothing culture. We believe this research work can not be accomplished overnight, but needs time accumulation and fully digestion. Teams always work with great care on the shape of the characters, trial and revise many times for the best elegance and composure of dress colors, and refine sentences for the accuracy and conciseness of words. Facts have proved that only after a long time of repeated learning and thinking, with experience and practice, we can have new discoveries and new inspirations, absorb and make good use of this treasure house of human art.

4. The significance and development direction of Dunhuang costume culture research

In 1900, the Dunhuang library cave was discovered, which shocked the world with its numerous and valuable treasures. Since the founding of the Dunhuang Academy in 1944, after several generations of scholars' efforts, Dunhuang studies have achieved rich research results and developed into a truly international subject. However, as far as the current research status of Dunhuang studies is concerned, most of them are focused on history, archaeology, World History, ethnology, linguistics, etc., and the special research on Dunhuang costumes is still scarce. Therefore, it is undoubtedly an important improvement for Dunhuang studies to comb and study the clothing culture in the murals of the past. In addition, from the perspective of the study of ancient Chinese clothing history, most of the records in China of clothing are official clothes and dress system of the Central Plains courts. There is still a lack of research on the clothing history and cultural connotation reflected along the Silk Road under the consider of different dynasties, ethnics and religions, and the research on the clothing culture of the common people and other nationalities. Therefore, the study of Dunhuang costume culture is also an indispensable part in the field of ancient Chinese costume culture research.

Scholars and designers in different times have different understandings and feelings about Dunhuang art. Contemporary scholars and designers should add cultural ideas, artistic thoughts and social fashion to the cognition of Dunhuang Art in the new era. In addition to the study, research and inheritance of Dunhuang clothing culture, in the context of the new era, the continuous development of social economy and the people's love and demand for traditional culture will undoubtedly put forward new topics for the clothing history researchers and designers. How to make the ancient Dunhuang costume art return to modern design and public life, and compose a more dynamic future? According to the research and experience of previous scholars, we know that using modern vision and design, drawing nutrition from the abundant Dunhuang costume art, endowing decorative elements with the sense of time and innovation, making the past serve the present, would fully display the cultural innovation and era charm of Dunhuang on the modern design stage and promote the traditional culture effectively. For example, Ms. Chang Shana created the silk scarf Dove of Peace as a gift for the Asia Pacific Regional Peace Conference in the 1950s. Later, she boldly integrated Dunhuang elements into the design of the ten major buildings in Beijing, and stood out among many design schemes. For more than half a century, these Dunhuang elements have gone through the test of time. Because of their magnificent and majestic style, they have been widely recognized and appreciated by the leaders and the public at home and abroad, reflecting the eternal vitality of Chinese traditional culture represented by Dunhuang art in the context of modern design.

This book is based on the existing study and research, focuses on re-examining, re-exploring and refining from the perspective of clothing culture, so as to obtain a new appearance and form of clothing, especially the cognition of clothing shape and structure, and the grasp of clothing pattern and color, which will generate new thinking and new interpretation. Wish the public could see: the long-standing Dunhuang art is moving towards the contemporary social life; Dunhuang art is not only classical culture, but also modern fashion, and it still has vitality in the new era. Therefore, through the publication and dissemination of this book, we hope that Dunhuang clothing culture can be understood, loved and used by more artists, designers and Dunhuang art lovers, and trigger more thinking about the combination of traditional culture and modern design. We should do some practical work in the construction of Chinese traditional clothing culture context and inheritance system, and guide the modern cultural and creative industry to produce more innovation and new design, so as to truly establish the cultural confidence of Chinese people and build the inclusive and grand Chinese spirit.

Liu Yuanfeng

Professor of Beijing Institute of Fashion Technology

Director of Dunhuang Costume Culture Research and

Innovation Design Center

January, 2021

序三

敦煌作为丝绸之路上的重要都会，自古以来就是不同民族、不同文化汇聚交流之处。营建于4～14世纪的敦煌莫高窟就是多元文化交融的产物，现存的45000多平方米壁画、2000多身彩塑展示着不同时代、不同民族的历史文化信息，成为当今诸多学科研究的重要资料。

敦煌壁画虽然是佛教壁画，但在表现佛像、佛国世界的同时，也描绘了大量的世俗人物及古代社会生活的场景，如其中的经变画、故事画、供养人画像等题材，真实地反映了历代社会生活的场面。画面中不同民族、不同身份的人物画像，他们的着装及相关装饰等方面，就是认识和研究古代服饰文化的历史资料。可以说敦煌壁画中各朝代服装的各类款式，以及服饰图案样式等，构成了一千年间的服饰文化史。其中不仅展现中国古代汉民族服装服饰发展历程，而且还反映了中古时期各少数民族以及外来民族的服饰样貌。仅就唐代而言，现存的270余个洞窟中展现出来的服饰如同一座座博物馆，从帝王贵胄到庶民百姓，从汉族绅士到各民族人物，那些官员、侍从、商人、农民、工匠等，展示着不同人物的服饰特色，如实地再现了大唐盛世的衣锦繁华。这些数量众多，内容浩繁的艺术形象，虽历经时光的洗礼和风沙的侵蚀，仍可看出古代服饰的诸多样式，甚至使用面料的质感，这丰富而完整的服饰艺术体系让人叹为观止。

长期以来，由于古代服饰实物资料的缺乏，中国服装史的研究存在很多困难。而现存各时期的出土文物及壁画艺术中的人物图像就显示出重要的资料价值。特别是南北朝至唐代服装的发展演变，可以借助敦煌壁画这样的丰富图像寻找其发展轨迹。自20世纪80年代以来，就有不少专家对敦煌艺术中的历代服装以及装饰图案进行探讨，著名学者沈从文先生以及段文杰、常沙娜、关友惠、谭蝉雪诸先生都先后从不同的角度对敦煌的服饰及装饰艺术进行挖掘和整理，发表过一系列论著，为后来的研究者奠定了基础。今天，服饰的研究不仅是对古代文化的探索，而且还要为当今的服装设计、装饰设计提供新的灵感来源。吸取中国古代传统文化中的有益成分，加以继承和发扬，正是艺术发展的必由之路。

2018年6月，敦煌服饰文化研究暨创新设计中心正式成立，在刘元风教授的率领下，敦煌服饰文化研究暨创新设计中心的研究人员不断深入调查敦煌壁画中的服饰内容，把科研和教学相结合，举办各类研讨会、培训班。同时积极开展创新设计，取得了一系列可喜的成果。以中心为依托，敦煌研究院与北京服装学院充分发挥两家单位的不同优势，在敦煌服饰文化领域进行理论研究、设计开发、人才培养等不同方面的深入合作，既开拓了敦煌学研究的新领域，也带动了敦煌服饰文化创新教学的发展。

《敦煌服饰文化图典》就是依托敦煌服饰文化研究暨创新设计中心申报的国家社科基金项目，是两院合作的重要内容之一，此项目旨在通过全面调查敦煌石窟，提取敦煌壁画各时期代表性的人物服饰以及相关装饰图案的典型作品，通过整理、绘制、复原，为学术研究提供系统的研究资料，为服装设计提供有价值的图像依据，也为广大敦煌艺术和服饰文化爱好者提供了解和欣赏敦煌服饰艺术的代表作品。因此，图典在选取敦煌壁画中代表性服饰图片的同时，通过研究和设计人员的临摹、绘制，突出壁画中服饰特色和装饰图案的风格。敦煌壁画经过千百年的风沙侵蚀和阳光曝晒以及各类病害，大多已变色或褪色，使相当多的人物形象变得模糊、残缺，不易辨识，研究人员通过对绘画的时代风格、服饰结构等方面的调查研究，采用复原的办法，使读者能够看到古代人物服饰的完整形象。这里需要特别说明的是，本研究进行的敦煌服饰复原有两层含义：一是以服饰效果图的形式展现——从服饰整体造型、结构特征、色彩配置、图案装饰等艺术形态进行绘画复原；二是根据复原效果图，选择相应的材料、剪裁方式及工艺制作方法，以成衣的方式进行复原。这也是我们此次对于敦煌服饰研究的特色与创新之处。相信通过这样的可视化形式将敦煌服饰文化向社会推广，不仅可以使相关行业人员从中找寻到可以进行开发利用、传承创新的资料，也可以使普通读者了解和欣赏敦煌服饰艺术，从而更进一步弘扬以敦煌石窟为代表的优秀传统文化。

敦煌服饰文化的研究作为探讨中国传统文化的重要领域，还有很多方面需要不断开拓，进行持之以恒的研究。而在继承传统文化的基础上进行服饰艺术的创新，创作出具有中国文化特色的新时代的服装艺术作品，更是任重道远。随着以敦煌艺术为代表的中国传统文化深入人心，弘扬敦煌服饰文化，创新设计敦煌风格服饰艺术，必将推动中国服装设计领域树立中国品牌，并走向世界。

趙聲良

敦煌研究院院长、研究员

2020年夏

Preface 3

As an important city on the Silk Road, Dunhuang has been the place where different nationalities and cultures gather and exchange since ancient times. Mogao Grottoes, built during the 4th–14th century, is the fruit of multi-cultural integration. The existing 45,000 square meters murals and more than 2,000 painted sculptures show the historical and cultural past of different times and different nationalities, they are important studying materials for many disciplines nowadays.

Although Dunhuang murals are Buddhist art, they also depict a large number of secular figures and scenes of ancient social life, such as in sutra illustrations, story paintings, and donor portraits. Some parts reflect clearly the social life in the past dynasties. The portraits have different nationalities and different identities, their clothes and decorations are best materials to understand and study the ancient costume culture. It can be said that the various styles and patterns of costumes in Dunhuang murals constitute a clothing culture history of a thousand years. They not only show the development process of Chinese Han nationality clothing, but also reflect the clothing style of various ethnic minorities and foreign nationalities in the middle ancient times. As far as the Tang Dynasty is concerned, the costumes displayed in the existing more than 270 caves are like many clothing museums. From emperors and nobles to common people, from Han gentry to ethnic figures, officials, attendants, merchants, farmers, craftsmen and so on, the clothing characteristics of different characters are vivid and clear, and truly manifested the prosperity of Tang Dynasty. These images are numerous in number and rich in content, although having experienced the baptism of time and the erosion of wind and sand, they are still bright in color, clear in shape, and we can even feel the texture of fabrics through the paintings. This rich and complete clothing art system is amazing.

For a long time, due to the lack of material evidence of ancient clothing, there are many difficulties in the study of ancient Chinese clothing history. The figures in the unearthed cultural relics and mural art in different periods have important data value. Especially from the Northern and Southern Dynasties to Tang Dynasty, the development process of clothing can be identified with the help of Dunhuang murals. Since 1980s, many scholars have discussed the costumes and decorative patterns in Dunhuang caves. Mr. Shen Congwen and Mr. Duan Wenjie, Ms. Chang Shana, Mr. Guan Youhui and Ms. Tan Chanxue have studied and sorted out the costumes and decoration art of Dunhuang from different perspective, and published a series of works, which laid the foundation for later researchers. Today, the study of clothing is not only for understanding the ancient culture, but also a new source of inspiration for today's fashion design and decoration design. It is necessary to absorb nutrition from ancient traditional art, inherit and carry them forward.

In June 2018, Dunhuang Costume Culture Research and Innovation Design Center was officially established. Under the leadership of professor Liu Yuanfeng, researchers of DCCRIDC continue to examine the clothing subject in Dunhuang murals, combining with scientific research and teaching project together, holding various seminars and training courses. At the same time, innovative design has been actively carried out and a series of fruitful achievements have been achieved. Based on the center, Dunhuang Research Academy and Beijing Institute of Fashion Technology have given full play to the different advantages of the two groups and carried out in-depth cooperation in theoretical research, design and development, personnel training and other aspects in the field of Dunhuang clothing culture, which not only opened up a new field of Dunhuang studies, but also led to the new development of innovative teaching of Dunhuang clothing culture.

The Illustration of Dunhuang Costume Culture is a national social science fund project declared by DCCRIDC. It is an important cooperation project between the two groups. The purpose of this project is to extract the representative costumes of different periods from Dunhuang murals and typical works of related decorative patterns through the comprehensive investigation, to provide systematic references for academic researchers by sorting, drawing and restoration. It provides valuable image resource for fashion design, and also includes representative works of Dunhuang clothing art for Dunhuang Art and clothing culture lovers. Besides selecting the representative clothing pictures, this book highlights the clothing characteristics and decorative patterns in the murals by copying and drawing. After thousands of years of wind and sand erosion, sun exposure and various diseases, some murals' color have changed or faded, making a considerable number of characters blurry, incomplete and difficult to identify. Through investigation and research on the painting style, clothing structure and other aspects, researchers restored them accordingly, so that readers can see the complete image of ancient people's clothes. What needs to be specially explained here is that the restoration of Dunhuang costumes in this study has two meanings: one is to display the details of the costume — the overall shape, structural characteristics, color configuration, pattern decoration and other artistic aspects of clothing; the other is to use the corresponding materials, cutting methods and production methods according to the restoration effect drawing, to make the real clothes. This is also the innovation of Dunhuang clothing research this time. It is believed that the promotion of Dunhuang costume culture to the society through such a visual form can not only enable the relevant industry personnel to find information that can be developed and utilized, inherit and innovate, but also enable ordinary readers to understand and appreciate Dunhuang clothing art, so as to further promote the traditional culture represented by Dunhuang Grottoes to more people.

As an important field to explore Chinese traditional culture, the study of Dunhuang costume culture needs continuous exploration and persistent research. It is a long way to go to create new clothing art works with Chinese cultural characteristics on the basis of inheriting the traditional culture. When the Chinese traditional culture represented by Dunhuang art is deeply rooted in the hearts of people, promoting the Dunhuang clothing culture and innovating the Dunhuang style clothing art deign will certainly promote Chinese brand popular in the world.

Zhao Shengliang

Researcher, President of Dunhuang Research Academy

Summer in 2020

目录

Contents

唐前期敦煌壁画的装饰艺术

赵声良　张春佳
敦煌研究院

唐代丝绸之路的繁荣，带来了社会经济和文化的快速发展，敦煌莫高窟在丝绸之路上的位置愈加重要，多元文化的进一步交流，佛教信仰的日益发展，促进了敦煌石窟的营建，壁画的绘制水平也达到了前所未有的新高度。敦煌的艺术家将来自印度、西域和中原长安等地的艺术风格相融合，在莫高窟中创作出前所未有的精彩作品，显示出了空前的艺术活力和高超的艺术水准。敦煌石窟大量的壁画和彩塑反映了那样一个时代的精美艺术，成为中国美术史上不可或缺的重要组成部分。

美学家宗白华先生说过："一个文化丰盛的时代，必能发明无数图案，装饰他们的物质背景，以美化他们的生活。"[1]莫高窟壁画虽然以佛像画、经变画为主，但分布于不同部位的装饰纹样同样也显示出艺术家们杰出的创造力。唐前期的莫高窟中，没有任何两个洞窟的装饰纹样是完全一致的，尤其是重点部位，如藻井——不同洞窟之间的藻井展现出来精彩纷呈的造型特征和色彩组合。装饰纹样虽然只是莫高窟艺术的一个很小的部分，但是它会随着洞窟整体风格的走向而展现出不同的特征，有些特征可能是对于传统样式的怀念，更多的则是新时期的创造。在大唐盛世文化交流与融合高度发达的时代，敦煌的装饰艺术也体现出蓬勃发展的旺盛创造力和生命力。

莫高窟唐代前期的洞窟分期，本文基本上按照樊锦诗先生的《莫高窟唐前期洞窟分期》[2]中对于敦煌莫高窟初唐和盛唐洞窟的时期划分，大体为四个时期：初唐大致为唐一期、二期洞窟，盛唐为唐三期、四期洞窟。

一、唐前期洞窟中装饰纹样的分布

（一）装饰纹样分布格局

莫高窟唐代前期的装饰，在洞窟中的基本分布状况，如图1所示。

如果以观览洞窟的视线顺序来审视不同建筑区域的"视觉效果重要性"，那么毫无疑问，在进入洞窟后，首先映入眼帘的是西壁佛龛，其中的塑像和相关装饰是全窟核心，其次是南北两壁，再次是东壁，窟顶对于全窟而言是相对概念性的意象空间。从装饰的分布来看，不论是哪个时期，藻井和佛龛内的背光、光头等都是装饰纹样比较集中的区域。但从唐代一期洞窟到四期洞窟，也是有一定的侧重点和思路的变化。唐一期洞窟中，大部分是沿着建筑立面的交界转折部位分布，如图第204窟，装饰带仿佛构架起洞窟建筑结构的框架形态；唐二期洞窟，如图1中第335窟，这样沿建筑立面转折部位分布的框架性装饰状态依然延续，但是基本是集中于窟顶；唐三期、四期洞窟中，西龛内集中了大量的团形纹样，窟顶的装饰更加趋向复杂，团形纹样在整体洞窟中占据了主导地位。

盛唐洞窟中，装饰纹样整体分布面积更大，并且除了框架边饰外更多地在塑像服饰、单体纹样等不

[1] 宗白华《略谈敦煌艺术的意义与价值》,《观察》周刊第5卷第4期，1948年。

[2] 樊锦诗《陇上学人文存·樊锦诗卷》，兰州：甘肃人民出版社，2014年。

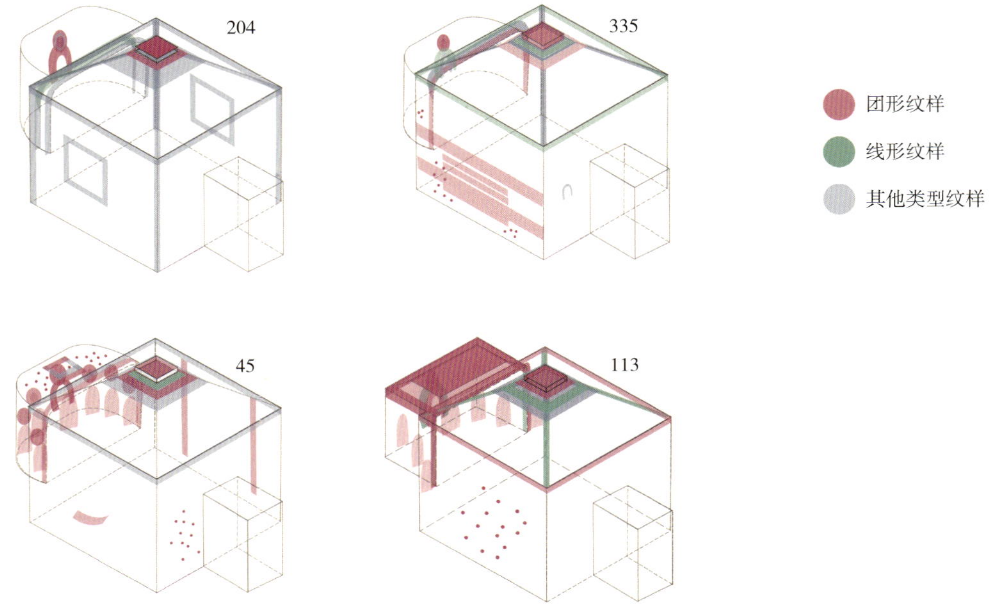

图1　唐前期洞窟装饰格局示意图

同方面加强使用团形纹样。这与盛唐社会愈加繁荣的整体趋势下，人们审美精神的变化息息相关。装饰纹样整体格局的变化是由局部细节逐渐演变来带动的。团形图案相比于线形图案，可以在单元面积内表现更多的细节元素。团花图案可以进行大量的复合——将多种题材的花卉和相关细节组合在一起；并且由于它是放射状中心对称结构，可以从中心点开始向外进行无限的层次叠加拓展，这是长线形纹样所不具备的特征，因而更为复杂华丽。

（二）装饰部位

唐代前期洞窟的形制基本上是以覆斗顶为主，少量中心柱窟，西壁开凿的佛龛大多为平顶，但是到了唐代四期洞窟有一部分盝顶帐形龛。洞窟的装饰根据洞窟形制而呈现不同的表现，装饰位置和面积也由窟内建筑立面划分决定，大致呈现在这样几个主要部位：

1．覆斗顶洞窟的藻井

这里是全窟装饰最为集中的区域，藻井的井心中有全窟面积最大、最富丽繁盛的装饰纹样，四披分布着层层叠叠的边饰，与井心共同模拟华盖的样貌。莫高窟早期的藻井仿照来自中亚的建筑结构，但是发展到唐代只留存这样的一个方井形式，内部模拟叠涩的结构已经去掉，同时加上周边四披的内容共同展现人们对于天国世界的美好想象。

2．边饰

唐前期的洞窟边饰主要分布在四披交界处、四披与四壁交界处、西龛内外边饰这样几处。大多为卷草纹或联珠纹装饰，少量出现几何纹或百花草装饰。大多用以区隔不同的绘画区域，强调建筑空间壁面转折分界。

3．背光

通常来讲，全窟较为复杂、绘制精美的头光和身光往往出现在西龛主尊佛和两侧菩萨、弟子背后。中心柱窟往往在中心柱正面和南北壁东侧有大型塑像，佛像、菩萨像的头后部也有塑出的头光，并有精美的头光图案。由于塑像体量较大，身后与之配合的背光也往往较大、较复杂。南北两壁和东壁的佛菩萨头光和身光由于壁画的面积所限往往较小，绘制的内容相对简单。背光，尤其是头光，由早期的火焰纹到唐代发展为纯植物纹样，但往往会有光的表现意味，用以表现佛常放光明之意。

4．其他部位的装饰

除了以上几处主要装饰部位，由于窟形不同，其所需要装饰的部位也有所区别，因而，以第332窟为代表的中心柱窟，前部人字披顶，其脊枋部分也有长条边饰。另外，唐四期盝顶帐形龛中，帐顶往往饰有平棋团花，平棋格边饰多为二方连续的小花边饰，平棋格内团花大多为六瓣的多层团花，虽不如藻井团花那般精美，却也比边饰或其他部位出现的团花要复杂得多。这些装饰部位由于建筑形制发生时代性的变化也会随之发生变化，但是都会在某种程度上丰富洞窟空间的层次感和视觉效果。

此处需要说明的是，还有很多装饰纹样绘于彩塑或壁画人物的服饰上，由于其分类较为独立，与洞窟壁画中的装饰纹样性质不同，本文暂不讨论。

二、装饰分类概况

唐前期纹样种类繁多，造型十分丰富。为便于把握唐前期装饰的总体发展情况，本文从结构上大致进行分类，主要的类型有两类：线形纹样和团形纹样。此外还有很多种类的装饰，因不占主导地位，笼统地归入“其他纹样”。

（一）线形纹样

北朝时期，忍冬纹题材极为兴盛，忍冬往往以波浪状二方连续，形成一种线形骨架，应用极为广泛。我们把这样的结构称为“线形纹样”。忍冬纹变化十分丰富，但线形结构是最为流行的。在北朝和隋代一直以忍冬或忍冬莲花的组合为主，及至初唐，缠枝卷草纹成了主流，多种花草形象融入卷草纹中，是唐前期装饰纹样的趋势（图2），特别是葡萄、石榴等纹样融合在卷草纹中，形成初唐时期的新气象，石榴卷草纹到盛唐时期更加流行。除了花草外，还有蜂、蝶甚至迦陵频伽等元素都常常融进了卷草纹中，形成唐代纹样的丰富特色。另外，忍冬纹并没有完全从唐代的装饰体系中消失，而是分解成不同的局部元素，融入各类装饰纹样中。在唐代线形纹样中，也有一些叶片的造型保留着忍冬的三裂和卷曲的造型，从形态上显示着对于旧日流行题材的继承。

线形纹样在唐前期基本骨架和整体纹样的外形为细长的线状或呈长条形，多为二方连续状，构成元素围绕着基本的线状骨架进行组织分布，多出现于各类边饰中。以石榴卷草为代表，整体呈长线状，骨架为二方连续的波浪状，或为多重波浪骨架。伴随这样的骨架，其中的构成元素可以是植物卷草，也可以是连续的几何纹样，叶片逐渐加肥，逐渐丰富。需要指出的是，线形纹样又可分为直线形和曲线形，曲线形多指头光等环状面积内的装饰纹样，但其基本骨架为曲线造型的卷草纹，因而归为线形。

（二）团形纹样

团形纹样也是在莫高窟最初就有，多以莲花的形式出现。北凉第272窟的藻井中心的团花就是单独的莲花，这样的团花形式在中心柱窟顶部的平棋图案和覆斗顶窟的藻井中不断出现，莲花的层次也不断丰富。同时在莲花周边的装饰也不断增加，到了隋代，往往会与飞天等形象共同出现于藻井中心。但莲花作为一个团形的装饰，没有太大的改变，基本维持莲花的正面俯视状。到了初唐时期，团花逐步脱离单一莲花题材，出现了其他多种花纹组成的团花形态（图3）。

团形纹样的外轮廓为圆形（传统纹样中的团窠，也可归入本文所说的团形纹样）。此类纹样为唐代最流行的纹样类别，以藻井井心团花为代表，多以植物纹样构成图案，常常呈十字或八瓣结构，盛唐时期出现六瓣，但多为八瓣结构。有时，根据布局的需要，也常常出现半团花纹样，其本质也属团形纹样类，因其单元结构的特征为团形。

图2　莫高窟第321窟龛内南壁 卷草纹 初唐

图3　莫高窟初唐第372窟藻井

（三）其他纹样

除了上述两类外，还有如菱形纹样、联珠纹样、几何纹样、龟背纹、方胜纹等多种纹样。

唐前期的几何纹样只少量保留，其中最主要的就是菱格纹。菱格纹往往以边饰形式出现，并且复合小花朵，也与菱格纹一致地呈一整二半之形，多分布于窟顶四披交界处、龛沿边饰部位、藻井边饰等。传自西亚地区的连心纹在莫高窟唐一期洞窟中多出现在佛龛边饰部位，但是进行了一定的“本土化”改造——心形图案变成三裂小花瓣，成为连瓣纹。此外，将莲花瓣、牡丹花瓣、卷须纹样等组合花朵，直接进行排列变成长边饰，也是唐代独有的装饰纹样。这些花瓣或纵向一整二半排列，或侧面分割成二分之一花瓣左右交错分布于长条装饰区域内。

还有一些装饰纹样有着特有的功能，如火焰纹，用以表现佛背光、头光；垂角纹、垂玲纹等用以表现藻井四边，象征着华盖的垂幔。

整体来看，虽然这一部分纹样细分类别很多，但是在唐前期装饰体系中的体量没有前面两个类型多，本文重在分析在唐代流行纹样中占主导地位的线形纹样和团形纹样，故对其他纹样暂不作详细分析。

三、线形纹样

莫高窟中的线形装饰纹样，多出现在各种边饰中。由于唐代洞窟大多为覆斗顶，西壁开龛，所以这些边饰多集中在藻井的四边、四披、四壁、西龛内外侧边缘以及造像的背光、头光等部位，另外还有一部分出现在造像的服装边饰上。这些边饰，依据所装饰面积的不同又可以大体分为直线形纹样和曲线形纹样。其中曲线形多集中在头光和背光边饰中，为环形和半环形区域装饰。

唐前期的线形纹样以卷草为主，主要题材为石榴，初唐出现的葡萄延续隋代的形式，只是局部偶尔绘制，后期并未大面积流行。

（一）骨架结构

唐前期的线形纹样多分布在长边饰部位，纹样骨架呈波浪状二方连续。但是这样的波状只是一种统称，自北朝发端的线形纹样，到初唐时期，骨架一直保持舒缓的起伏状态，如图4所示第220窟西龛龛楣，但是盛唐时期，大量出现单元翻卷近乎圆形的线形纹样，如第217、172、444窟。一改早日的舒朗修长，力图在单元图形内使花朵和枝叶的卷曲工整，并努力保持纹样单元间的一致性。这样的规则、严谨的骨架，与来自中亚和印度的影响有关。

初唐时期卷草纹的骨架多为单线骨架，骨架单纯为波浪形，所有的花朵和叶片生长在骨架之上，但是并不与骨架发生交叠，而是完好地避开骨架曲线，使骨架曲线可以非常清晰地呈现出来。盛唐时期，卷草纹小分枝会与骨架穿插遮挡，并且出现了大量复合的多重藤蔓骨架的卷草纹，诸多洞窟在边饰的设计上都采用了这种方式（图5）。这样的多重骨架，其源头可以上溯至战国时期的线形装饰纹样。而到了

图4　莫高窟初唐第220窟龛楣石榴卷草纹

图5　莫高窟盛唐第444窟龛边饰石榴卷草纹

唐代，变得更加丰富、厚重了。

唐代的头光中经常出现曲线形的卷草纹样，这类纹样围绕着头光中心进行排列，花叶在此基础上分布于空隙之内，如第208、217、41、444等窟西龛内头光边饰，多为石榴卷草（图6）。分布于头光环状区域内的卷草纹，其单元造型很容易使人联想到印度笈多时期的头光样式，但是笈多风格的头光纹样，基本上是以静态的花纹分布，敦煌此时的卷草纹则是通过波状的曲线相连，构成一种流动的趋向。

如果考察中国传统的线性造型，在新石器时代晚期，陶器装饰纹样中就可以见到大量的涡卷、水波等线造型装饰。这样的波状线在构成装饰效果、填充装饰表面和相关区域时有着非常强的适应性，造型较为随意并且具有富于流动性的生命力。秦汉时期，尤其是汉代出土的大量漆器，其表面的装饰纹样基本是长线造型，构成云气和动物等不同形象，都力求舒朗灵动（图7）。如图8所示西汉时期出土的漆绘陶钫上，有流畅的云气纹，用一簇一簇的线表现云气的流动感，造型流畅优美，这类纹样造型对于后世

图6　莫高窟盛唐第217窟头光

图7　湖北云梦睡虎地出土秦代漆盘

图8　西汉漆绘陶钫

的曲线尤其是线的生动性有着深刻的影响。中国南朝众多画像砖利用长波状线造型塑造出云气、衣饰、龙、植物等多种形象。从汉代到六朝的纹样中，我们看到这些舒展而流畅的云气纹或其他相关的纹样，往往构成一种动态的空间，因此，并不会把纹样充满绘画区域，而一定留有大量的空隙，形成特定的空间布局，表现出空灵的气韵。这种构成形式成为相当长时期的一个传统。在初唐时期莫高窟的装饰纹样中可以看出延续着始于西魏时期的舒朗的线造型，并将局部细节具体化，比如融入多种植物题材，形成了初唐时期特有的卷草纹样造型特征。这些以卷草为主的纹样很快走向饱满化，其装饰面积中所留的空间越来越少，纹样与背景的关系与早期差别很大，中晚唐延续盛唐的饱满态势，纹样填充几乎无空隙。从初唐到盛唐时期卷草纹的变化中，可以看出由汉代空灵、飘逸的装饰风格正在急速地向复杂、饱满发展的趋向，而由线形纹样构成的活泼的动态，则仍然在保持。

（二）局部造型特征案例

1．忍冬纹的演变

忍冬纹在北朝多以线形纹样出现，在初唐时期的线形纹样中，也可以见到忍冬的痕迹。例如，第329窟藻井边饰（图9），是首尾相连的卷草纹样，主体的骨架可以看出波浪形的枝干，其中翻卷的枝叶虽然明显带有唐朝卷草的丰富特色，但其主干的叶子仍可看出北朝忍冬纹的样貌。只是在唐代壁画中，忍冬纹在造型上有了很多变化——叶片变得更为肥厚，尖端变得圆润并且会有回卷，忍冬叶片朝着卷须的方向在发展并嵌入牡丹花瓣，将整体改造成花苞样造型，似乎成为忍冬向唐代卷草和团花细节演变的例证。这里的卷草纹样所进行的尝试使人们看到了纹样在发展过程中的演变过程，某些造型组合在萌芽之后得到发展，而另一些则消失不见。

2．百花草与小花朵边饰

百花草边饰是众多线形纹样中较为特殊的一类，在盛唐时期洞窟中局部流行。它并不同于通常意义上的二方连续的卷草纹，因为没有特别明确的曲线骨架，也很难界定它的纹样循环单元。并且纵然是整条纹样大体为二方连续，但实质上每个单元都不同——无论是纹样的种类或造型都很难找到完整确切的循环单元。目前，盛唐洞窟保留下来的肉眼可见的壁画中大约有十五个左右的洞窟中出现百花草边饰（图10）。这些边饰分布在四披交界处的边饰上以及龛沿边饰、头光边饰、藻井边饰等部位（图11）。每一个单元局部的花朵和叶片都尽可能有所差别，使得整体本就多变的花草看起来更加丰富多彩。如果说石榴纹卷草或许因为一定的粉本原因使得每个循环单元趋于一致，那么，百花草纹样可以称得上是展示画家创造力的舞台：大面积的左右对称中饱含变化，任何两个局部都不相同，画家会极尽所能地展示艺

图9　莫高窟初唐第329窟藻井

图10　莫高窟盛唐第66窟百花草边饰

术的创造性和差异性，这种纹样从一定程度上来讲达到了线形纹样复杂化的顶峰。

初唐和盛唐时期的装饰纹样，以藻井为例，没有任何两个洞窟完全一致，甚至不会使用看起来较为相似的粉本。而百花草于小小的面积上集中了大量形态各异的构成元素，可以从一管窥全豹——百花草纹样从某种意义上代表了盛唐洞窟对于装饰和审美的理想表达，说明盛唐洞窟装饰蕴含着无比丰满的创造力和不断追求创新的审美精神，这是唐代前期一种审美理想的缩影。

四、团形纹样

莫高窟唐前期洞窟壁面和彩塑装饰中，团花是最为重要的纹样之一。这与团花纹样在当时社会中广泛流行相一致。唐前期的团花纹样以四瓣的十字结构和八瓣的米字结构为主。初唐团花的基本结构由四个主要花瓣构成，并且依据这一十字结构进行装饰。从汉代铜镜中的十字结构的柿蒂纹来看，其与唐前期流行的十字结构的团花从造型上有一定的渊源关系（图12）。无论从唐初对十字结构的偏爱还是单体莲花瓣的造型特征来看，二者都有联系。盛唐时期的团花纹样占据主导的是八瓣结构，并且出现更多的花瓣数量结构。由初唐至盛唐的内部变化，从骨架上看就是从十字结构向米字结构的转变，而花瓣的层次则由单层向多层发展。

团花纹样于初唐时期开始形成真正意义上的“团花”的造型——复合不同的植物纹样题材，共同构成俯视的多层次花朵，并且会以最为华丽的复合形态装饰于窟顶的藻井井心中，以这一个核心点带动全窟的装饰。佛造像是洞窟开凿营建的重点，是礼拜的视觉核心，但是洞窟的装饰核心在覆斗顶洞窟中却是集中于藻井的——这里出现的团花或层层叠叠的边饰是全窟的装饰重心，窟顶四披交界处的边饰以及其他部位装饰（在造型、色彩特征等方面），都从属于井心的艺术风格。

（一）局部造型特征和主要题材

初唐时期，团花的造型有少量还保留着隋末的痕迹——也就是藻井中单纯俯视莲花的造型特征：四角为四分之一莲花，整体几乎没有层次的区别，都是双层花瓣。很快莲花的结构就开始复杂化并且复合各种题材。当然，融合进来的不同题材中，得以长久地保留的就是牡丹花和卷须。牡丹加卷须和侧卷瓣莲花共同形成了唐代团花尤其是唐代前期团花的最主要的识别特征，与此同时，卷须、三裂牡丹花瓣、如意纹、石榴、葡萄等题材悉数登场（图13）。

图11　莫高窟盛唐第225窟百花草头光边饰

图12　西汉“中国大宁”瑞兽博局纹鎏金铜镜

图13 莫高窟初唐第387窟、第205窟

越来越多的牡丹、石榴、茶花等当时流行的花卉出现在莫高窟的装饰纹样中。显然它们并不具备佛教艺术中的那些象征意义，而是宗教美术与世俗生活融合的表现。当然，莲花的主导地位并没有改变，但是除了用于部分佛像头光的中心部分，以及脚下的莲花之外，其他地方出现的莲花装饰，也已不再重视宗教意义的表达，而更多倾向于单纯的审美性再现。而莲花以外大量世俗社会中流行的装饰纹样，被应用在唐代洞窟的装饰中。

盛唐时期与初唐时期相比，大部分的结构是延续下来了，唐代三期洞窟中，很多团花图案的花瓣被复杂化，也就是尽量丰富地装饰各种元素；然而到了唐代四期洞窟中，也就是盛唐最后一个时期，艺术家们已经不满足于将花瓣仅作为“花瓣”来丰富化处理，他们将花瓣的部分直接绘制成小花朵，使用小花朵构成大团花的一个层次。这样的状态对于装饰的整体发展而言，依然是纹样尽量丰富化的一个步骤，这样的手法在唐代四期达到了巅峰，并且再无进一步的复杂形态出现（图14、图15）。中唐时期，单独花朵的样貌倒是保留下来了，只是整体都趋于简单明了，如第201窟中出现以茶花为组成元素的团花时，

图14 莫高窟盛唐第444窟团花头光

图15 莫高窟盛唐第185窟藻井（线图）

也可以看到这样的尝试被延续和继承下来，并且在晚唐时期表现为花环式的团花，也就是团花纹样是由6～8朵小花组成，彼时的创造力倒是在大体一致的艺术风格中下降了，装饰纹样也愈加简单和单一。

（二）半团花

半团花往往不单独出现，而是成组地构成边饰。但是在藻井井心立面上可以单独占有一个装饰面。在边饰中，较常出现的一种样式就是“一整二半”，也有学者称之为“一整二破”，意指在这样的团花边饰中，较大的完整团花排列在中央区域，两侧的空隙里间错分布着半团花（图16）。这些半团花和团花组成的循环单元中，是由一个团花和两个半团花组成的，因此称作一整二半；或者都是由半团花交错组成，如第387、335、340、341、321、320、79藻井边饰等。半团花边饰在唐代前期洞窟中大量出现，可以出现在任何边饰部位，比卷草纹的应用更为广泛，且更具代表性，尤其是在服装边饰上，往往不使用卷草纹，而使用半团花。在这里半团花已具有了线形纹样的特点。

图16　莫高窟盛唐第217窟藻井边饰

这样的线形装饰格局可以拓展成装饰面，并且在唐代纺织品中可以经常见到这样的团花纹织锦。只是大部分的织锦纹样，均有主团花和辅助花，主团花形态饱满，层次丰富，多为八瓣的米字结构，而辅助花填充在空隙处，多为十字结构的小团花，其边缘也多向内凹，以适应主团花饱满的团状轮廓，从而使两者之间所围合而成的空间是和谐流畅的环状。

半团花边饰中有一类较为极端的例子，就是半团花的设计只是考虑作为半花出现时的效果，其主体的思路并没有将其复原的意愿，因而，它们的结构设计并不能被完全复原成团花，这类案例可见于第41、31、79窟的窟顶藻井边饰等处（图17）。

图17　莫高窟盛唐第31窟藻井边饰

（三）类团花

还有一类装饰纹样，其外轮廓总体呈圆形，但是并没有完全按照放射状中心对称的格局，也看不出十字结构、六瓣或八瓣，而往往还会有侧面花的形象，甚至会有半团花组成的边饰、卷草等，但其外轮廓保持团状，内部往往以不同的结构来填充（图18）。总体的装饰面积形状为团形，因而称为类团花。这类纹样多见于洞窟壁画中的头光、身光等部分。

（四）联珠团花

联珠纹是北朝到隋唐时期十分流行的纹样，源于波斯的纺织品纹样，其基本构成是由如珠子般的小圆形组成圆环，圆环内往往描绘动物或狩猎纹样。考古发现的北朝纺织品中较常见（有的可能就是由波

图18　莫高窟盛唐第217窟头光

斯传入的）。在莫高窟壁画中于隋代中期正式出现，莫高窟第292、425、402、401、390、394、56等窟中均有联珠纹出现，有的是直接按外来的形式，表现狩猎或动物图案的联珠纹，也有的以花卉图案绘于联珠纹中，如第425窟西龛边饰的联珠纹中，可以见到清晰的十字结构团花纹样，这种装饰的样式在唐代一期洞窟中被延续下来。因联珠纹也是以团形表现的，在本文中把它归入团花类。

团形联珠纹在唐代同时期工艺品中出现较多，环形联珠围绕的中央往往是狩猎或对羊、对鹿等动物纹样或植物纹样。莫高窟隋朝开始大量出现联珠纹并延续到初唐，其后，就很少见到团花形的圆环联珠纹了，较多的洞窟采用单纯的圆形组成的一字排开的线形联珠纹。这些单纯的联珠作为边饰装饰在洞窟中多种建筑面交界处，如四披交界处、西龛边饰等处，或者用以丰富多层次线形纹样的装饰面，如藻井边饰。在隋代洞窟的边饰中大量出现的联珠纹，从某种意义上辅助着团形忍冬纹共同将装饰纹样整体的发展趋势从线形转向团形，也就是单元分割化更加明确。而西域联珠纹以狩猎纹、有翼兽等为主题的图像逐渐被植物纹样取代，在洞窟的边饰中呈现出更为平和稳定的视觉效果。

五、色彩和技法

装饰纹样的色彩构成与洞窟的整体色彩布局是一致的。北朝壁画中西域式晕染法流行，注重人物的立体表现，装饰纹样的层次也较丰富，往往以叠晕法描绘忍冬纹等纹样，显得厚重而丰富。北周到隋代壁画中较为重视线描的作用，更多地体现出平面的装饰感，如第292、420、207窟等，其中的忍冬纹很多都是以白色或黑色的线来造型，至于内部填充的色彩似乎可有可无，纹样之所以存在流畅飘逸的视觉效果，更多是因为线条的造型组织在起作用。初唐时期，由于中原的新风格大范围地影响莫高窟壁画的绘制，以第220窟为代表的新时期壁画中，线条的独立生命力愈加受到重视，并且设色与线条结合紧密，颜色从很大程度上也在展现着形体的起伏变化，进行微妙细腻的衔接和晕染。这些精妙的细节是前朝画家从技法上所难实现的效果。初唐色彩较为典型的品种有：石绿、石青、赭石、浅蓝、墨绿、土红。初唐有的洞窟保留了隋代的大面积的土红和群青色的组合，但在唐代以后的进一步发展中，洞窟的主体色调以明净而绚丽的青绿色调为主，装饰图案也大多是以石青、石绿的组合，辅以土红色、白色等元素。由于部分颜色的氧化变色，变成了棕黑色，形成了今天我们所看到效果。但透过这些变色，我们仍然可以感受到初唐装饰的绚丽、清爽的氛围。莫高窟外部环境地处荒漠，植被覆盖极少，古代的朝拜者经过漫山遍野的沙土色，进入洞窟后所看到的青绿色彩，所感受到的佛国世界的绚丽斑斓，定会在他们心中产生极为震撼的情感。从这一角度来讲，初唐洞窟营建的艺术家们非常善于营造一种独特的宗教色彩氛围。

盛唐时期的洞窟，总体色彩特征是在初唐的基础上进一步加强色彩的丰富与浓丽，石绿色、石青色与红色的组合也继续在洞窟中占据主导，但以土红、朱砂为主的红色调较以前大幅增加了。装饰纹样造型与初唐相比，更加饱满复杂，色彩中的石青、石绿、中绿、大红、土红和白色以及金色相配合，形成富丽堂皇、仪态万千的景象。由于氧化变色，壁画中存在大量的黑色、棕色，形成我们现在所见的效果。盛唐洞窟，尤其是第四期，线条和色彩的细腻程度比初唐大多数洞窟要更加精致入微。以第172窟为例，壁画中的诸多细节可远观，也可近处仔细欣赏——画家对于整体效果和细节的把控都非常重视。如此一来，对于朝拜者或观者，其长时间驻留洞窟的话，也会有几乎无尽的细节和禁得起推敲的局部来供其慢慢欣赏。

小结

莫高窟从北朝洞窟开始，装饰纹样呈现出以线造型为主导的重要特征，十余种不同类型的线形忍冬纹样构成了北朝时期洞窟装饰的主体，而同时忍冬纹本身就具有流畅的线造型特征。隋代是个承前启后的时期，长线形忍冬花是对忍冬纹的一种单元式分解，同时也是对线造型的继承。初唐时期继承了北朝至隋代线形纹样的形式，但在细部的表现上已有了很大的改变，特别是以卷曲多样而层次丰富的卷草纹代替了早期以忍冬为主的形式，带来了新颖的活力。初唐的众多卷草纹或类团花纹样的构成方面，主体骨架都较为舒朗，造型元素十分强调“线”的流畅动势。无论植物的种类如何，都是以线造型为主，线造型决定了纹样的疏密、细节的扭转变化等，色彩配合主干的线条，增强了丰富性。上溯至秦汉时期，就可以看到在中国传统装饰和绘画艺术中，线造型的重要性是无可比拟的，通过线的灵动、变化，营造出一种充满生机的气息，就是所谓“气韵生动”境界。这样的传统一直持续地影响着后世的艺术创作。唐代的画家们将装饰纹样中的线赋予一定的内在审美精神，使其具有西域的纹样中的线所不具备的独立生命力，向内追求其精神性的表达，从而使装饰纹样的枝叶造型也符合本土的时代审美特征，这也正是唐代画家们充沛创造力的具体表现之一。

从隋朝开始，众多边饰中都出现了团形忍冬——也就是由忍冬叶片旋转或以十字结构组合成团形忍冬纹，并构成长边饰。忍冬纹由纯线形向具有单元化区分的团形纹样转变可以看作是线形纹样向团形纹样转变过程中的一种尝试。同时，将忍冬纹本身进行一定的分解和转化，融入唐前期的团花和卷草的局部细节中。此外，来自波斯地区的联珠纹的流行也从某种角度上为线形纹样向团形纹样的流行转变推波助澜——联珠纹本身具有团形的特征，同时闭合的边界形成环状的区域，其多以二方连续构成边饰。这样的团形纹样在大量盛行于唐代工艺美术品上之时，也在莫高窟的装饰纹样中表现出同样的倾向性。唐前期，含有团形元素的边饰与卷草纹并存的同时逐步占据主导优势，并且以藻井团花为引领，整体洞窟装饰呈现出团形图案的强盛势态。石窟装饰与世俗工艺美术存在千丝万缕的联系。相关工艺水准的大幅提升对壁画中的相关装饰形式也产生了重要影响，敦煌的装饰同样是唐代前期趋于繁盛华丽的社会精神的具体表现。团形，具有丰满、富足、繁盛的气象，是大唐盛世的象征。以团形为主的装饰纹样与同时期的纺织品、金银器皿关系极为密切，表现出十分相似的造型特征，也包含了来自多个文化发源地的源头因素，包括西亚、印度、中亚、长江流域等地，是农耕文化的高度发达和社会文化的繁荣景象的综合展现。这些源自异域的风格在初唐和盛唐时期又不断地与本土的艺术表现和审美倾向相融合。

唐朝的大一统对社会的稳定发展和经济上升起到很重要的基础性作用，文化和艺术也趋向繁荣稳定。唐代寺观壁画十分兴盛，有大量优秀画家从事佛教壁画创作，如《历代名画记》中提及的进行过寺观壁画或相关佛教绘画创作的画家就有33人，此外，如《酉阳杂俎·寺塔记》也记录了29位画家，可知当时的画家主要作品都是在寺观之中。寺观与人们的生活密切相关，是文化生活的重要活动场所。及至敦煌，从长安涌入的大量画家带来了中原的高超技艺和多样的艺术风格。中原的艺术创作，以首都为中心对周边的影响十分强大，这样的艺术风格涉及了社会的各个领域，寺观壁画的创作也同样，在这样一种社会风气和思想观念影响下，呈现出独特的创造性。

从线形纹样为主导到团形纹样为主导，这是一个装饰思路的转变过程，也可以说是一种艺术表现风格的演变过程，意味着单元结构的日趋复杂化。莫高窟装饰艺术作为唐代艺术的局部案例，无论从题材、造型、格局还是色彩方面都是由单纯向复合、简单向复杂、疏朗向茂盛方向演变的，多种题材花卉的融合应用、花朵构成元素的日益复杂化都从不同方面展现了装饰纹样的整体发展趋势——呈现出愈加浓郁、繁复华丽的态势，这与唐代艺术、文学乃至社会生活的走向都是一致的。在这样的整体文化环境下，莫高窟的装饰艺术呈现出趋向繁盛的风貌。

Decorative Art of Dunhuang Murals in the Early Tang Dynasty

Zhao Shengliang & Zhang Chunjia
Dunhuang Research Academy

The prosperity of the Silk Road in the Tang Dynasty brought the rapid development of social economy and culture, and the position of Dunhuang Mogao Grottoes became more and more important on the Silk Road. The further communication of multi-culture and the growing development of Buddhist belief promoted the construction of Dunhuang Grottoes, and the painting level of murals also reached an unprecedented new height. The Dunhuang artists combined the artistic styles from India, the Western Regions and Chang'an of the Central Plains to create unprecedented wonderful works in Mogao Grottoes, showing unprecedented artistic vitality and developed artistic standards. The large number of murals and colored sculptures in Dunhuang grottoes reflected exquisite art of that era and have become an indispensable part of Chinese art history.

Mr. Zong Baihua, an aesthetician, once said, "In an era of rich culture, numerous patterns will be invented to decorate their material backgrounds and beautify their lives."[1] Although the murals in Mogao Grottoes are mainly paintings of Buddha images and sutra illustrations, the decorative patterns distributed in different parts in caves also show the artists' outstanding creativity. In the Mogao Grottoes of the early Tang Dynasty, no two caves have exactly same decorative patterns, especially the key parts, such as caisson — caisson among different caves show wonderful modeling features and color combinations. Although the decorative pattern is only a small part of art in Mogao Grottoes, it also follows the overall style trend of the grottoes with different characteristics. Some characteristics may be the nostalgia for the traditional style, while more are the creation for the new era. In the prosperous period of the Tang Dynasty, when cultural exchange and integration were highly developed, Dunhuang's decorative arts also showed vigorous creativity and vitality.

In this paper, the grotto's periodization in the Early Tang Period of Mogao Grottoes is basically based on Fan Jinshi's *Grottoes Periodization in the Early Tang Period of Mogao Grottoes*[2]（《莫高窟唐前期洞窟分期》）In the early Tang Dynasty and the high Tang Dynasty caves of Dunhuang Mogao Grottoes are divided into four periods: the early Tang Dynasty is roughly the first and second phases, and the high Tang Dynasty is the third and fourth phases.

---

❶ Zong Baihua（宗白华）, A Brief Discussion on the Significance and Value of Dunhuang Art（《略谈敦煌艺术的意义与价值》）, Observation Weekly《观察》, Vol. 5, Issue 4, 1948.

❷ Fan Jinshi（樊锦诗）, Long shang xue ren wen cun, Fan Jinshi Volume（《陇上学人文存・樊锦诗卷》）, Lanzhou: Gansu People's Publishing House, 2014.

1. The Distribution of Decorative Patterns in the Early Tang Dynasty Caves

1.1 Decoration pattern's position in caves

The basic distribution of decorative pattern in Mogao Grottoes during the early Tang Dynasty is shown in Fig. 1.

If we use a visitor's perspective to observe caves in order to examine the different areas' "visual effects significance" of a cave, there is no doubt that after entering the cave, the first thing you encounter is a west wall niche, statues and related decoration are core elements in a cave, then north and south walls and east wall. Ceiling to a cave is a relative conceptual space. By analyzing the distribution of decoration, we know that no matter which period, caisson, backlight and head light inside niche etc., are the area with more intensive decorative patterns. However, from the first phase of the Tang Dynasty caves to the fourth phase caves, there are certain changes in emphasis and thinking. In the first-phase caves of Tang Dynasty, most of them are distributed along the junction turning point in caves. As shown in Cave 204, the decorative belt seems to frame the architectural structure of the cave. For Cave 335 of the second period of Tang Dynasty, as shown in Fig. 1, such frame decoration along the turning points of the structure still continues, but it is basically concentrated on the top of the cave. In the caves of the third and fourth periods of the Tang Dynasty, a large number of round patterns were popular in the west niche, and the decoration on the top of the caves tended to be more complex, and the round pattern dominated the whole caves.

In the high Tang Dynasty grottoes, the overall distribution area of decorative patterns is larger, and besides the frame decoration, round patterns were used more intensively in other places such as statue clothes and monomer patterns. This is closely related to the change of people's aesthetic taste under the overall trend of the prosperous society in the high Tang Dynasty. The change of the whole pattern style evolves gradually by specific details change. Round pattern can present more details in a unit area than linear patterns. Round pattern can be very complicated — combining flowers of various themes and related details together; and because it is a radial

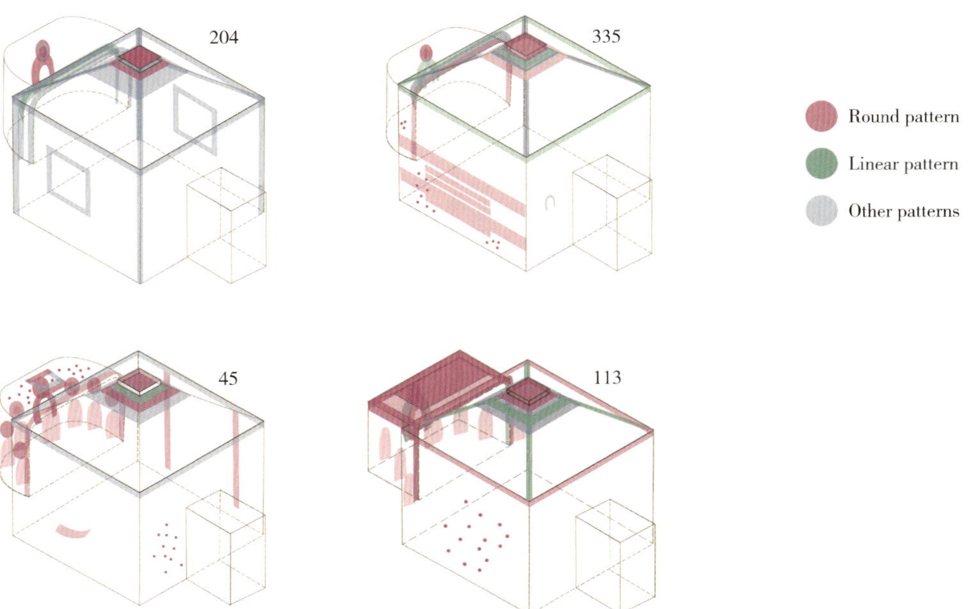

Fig. 1 Schematic diagram of cave decoration patterns in the early Tang Dynasty

centrally symmetric structure, it can carry out infinite layers superposition and expansion from the center point outwards, which is a feature that linear patterns do not have, so it is more complex and gorgeous.

1.2 Decoration position

The early Tang Dynasty grottoes architectural structure are mainly truncated pyramidal ceiling cave, as well as a small amount of central column cave. The west wall niche usually has a flat top, but some of the fourth period Tang Dynasty grottoes have tent-shaped niche. The decoration performance changes according to the shape and structure of caves. The decoration area and location are also determined by the surface division in the cave, which are generally shown in the following places:

1.2.1 The caisson on the top of the truncated pyramidal ceiling cave

This is the most intensive area of decoration in the whole cave. The center of the caisson has the largest area and the richest and the most prosperous decorative patterns in the whole cave. The four slopes have many layers edge decoration together with caisson center to simulate the appearance of a canopy. In the early stage, the caisson in Mogao Grottoes imitated the architectural structure from Central Asia, but when it developed into the Tang Dynasty style, only such a square well form was retained, and the internal simulation of overlapping structure was removed. Meanwhile, the contents on the surrounding four slopes were added to show people's beautiful imagination of the celestial world.

1.2.2 Edge decoration

Cave edge decoration in early Tang Dynasty is mainly distributed at the junctions of the four slopes, four slopes and the junctions of four walls, and the interior and exterior edges of the west niche. Mostly are scrolling grass pattern and Sassanian roundels pattern, and a few are geometric pattern or floral pattern. Most of them are used to distinguish different painting areas and emphasize the turning boundary of the architectural space.

1.2.3 The backlight

Generally speaking, the complexly and exquisitely painted head and body lights in a grotto often appear behind the main Buddha, Bodhisattvas and disciples on both sides of the west niche. There are often large statues in the front of the central column and on the east side of the north and south walls in the central column grottoes. There are also sculpted head light behind the heads of Buddha and Bodhisattva statues, as well as exquisite head light patterns. Due to the large size of statues, the backlight behind also large and complex. The head light and body light of Buddha and Bodhisattva on the north and south walls and the east wall are usually smaller due to the limited area of the murals so the content of the painting is relatively simple. The backlight, especially the head light, developed from the early flame pattern to pure plant pattern in the Tang Dynasty, but often had the expression of light, to show the meaning that Buddha always emitting light.

1.2.4 Decoration of the rest areas

Except the main decorative parts mentioned above, due to the different grottoes structure, the areas that need to be decorated are also different. Therefore, the central pillar grottoes represented by cave 332 have a gabled ceiling in front part of the main chamber, the ridge and beam parts also have long edge decoration. In addition, in the tent-shaped niche of the fourth period Tang Dynasty caves, the niche ceiling often decorated with flat chess flower medallion pattern. The flat chess lattice decoration usually are floret edges, flower medallions in lattices are usually six petals multi-layered, not as exquisite as caisson flower medallion, but more complex than edges or other areas flower medallion. These decorative parts will also change due to the changes of architectural form in different times, but all of them have enriched the sense of depth and visual effect of the cave space to some extent.

What needs to be noted here is that there are many decorative patterns painted on clothes of colored sculptures or figures in murals. Because their classification is relatively independent and different from that of the decorative patterns in cave murals, this paper will not discuss them here.

2. Decoration Classification Overview

There are many kinds of patterns in the early Tang Dynasty, and their shapes are very rich. In order to grasp the overall development of decoration in the early Tang Dynasty, this paper roughly classified it from the perspective of structure. There are two main types: linear pattern and round pattern. In addition, there are many other kinds of decorations, and because they are few and secondary, we classify them roughly in "other types".

2.1 Linear pattern

During the Northern Dynasty, the honeysuckle pattern was very popular. The honeysuckle pattern is usually two as a group repeated, forming a linear frame. It was widely used. We call such a structure a linear pattern. Honeysuckle pattern varied widely, but the linear structure was the most popular. In Northern and Sui Dynasty mainly were honeysuckle, or a combination of honeysuckle and lotus pattern, then to the early Tang Dynasty, tangled vine scrolling grass pattern had become the mainstream, and a variety of flowers and plants were added into the scrolling grass pattern. It was the trend of the early Tang Dynasty decoration pattern (Fig. 2), especially grape, pomegranate and so on were added into the scrolling grass pattern, offering a new style for the early Tang Dynasty. Pomegranate scrolling grass pattern became more popular in the high Tang Dynasty. In addition to flowers and plants, there are bee, butterfly and even Kalavinka and other elements being often integrated into the scrolling grass pattern, forming the rich characteristics of the Tang Dynasty pattern. On the other hand, honeysuckle pattern did not completely disappear from the decorative system in the Tang Dynasty, but was broken into different partial elements and integrated into various decorative patterns. In the linear pattern of the Tang Dynasty, there are also some leaves that retain the shape of honeysuckle's triple splits and curves, which shows the inheritance of the old popular theme form.

The basic frame and the overall pattern of linear pattern in the early Tang Dynasty were slender and linear or elongated in shape, and most of them were two in a group repeated. The constituent elements were organized and distributed around the basic linear frame, and most of them appeared in various trimmings. Represented by the pomegranate scrolling grass pattern, the whole pattern is a long stripe, the frame is two in a group repeated waves, or multiple wavy structure. Along with such frame, of which the composition of the elements can be vegetation scrolling grass, or continuous geometric patterns, and the leaves gradually become plum and complex. It needs to be pointed out that the linear pattern can be divided into straight and curved. Curved linear pattern is more common in head light and other circular area of decoration, but the basic frame is curve-shaped scrolling grass pattern, therefore it is classified as linear pattern.

2.2 Round pattern

The round pattern was also found at the beginning of Mogao Grottoes, mostly in the form of lotus flowers. The flower medallion in the center of cave 272 caisson dated to Northern Liang Dynasty is a single lotus flower.

Such flower medallion style constantly appeared in the flat chess pattern on the top of the central pillar cave and on the truncated pyramidal ceiling cave's caisson, and the layers of the lotus flower were constantly enriched. At the same time, the decoration around the lotus also increased constantly. In the Sui Dynasty, the images of flying Apsaras often appeared in the caisson center. But the lotus, as a round-shaped decoration, had not changed much. They basically maintained the lotus overlooking shape. In the early Tang Dynasty, flower medallion gradually separated from the single lotus theme and different themes combined into a round pattern emerged (Fig. 3).

The outer contour of the round pattern is round［the nest pattern（团窠）in the traditional pattern can also be classified as the round pattern in this paper］. This kind of pattern is the most popular pattern in the Tang Dynasty, represented by the flower medallion in the center of caisson. Most of the patterns are made up of plant elements, often in the shape of a cross or eight-petaled structure. In the high Tang Dynasty, six petals structure appeared, but most of them were eight-petaled structure. Sometimes, according to the needs of the layout, there are also half flower medallion pattern, but its nature also belongs to the round pattern, because its unit structure is round.

Fig. 2 Scrolling grass pattern on the south wall of cave 321 in Mogao Grottoes dated to the early Tang Dynasty

Fig. 3 Caisson in Cave 372 of Mogao Grottoes dated to the early Tang Dynasty

2.3 Other patterns

In addition to the two categories above, there are a variety of other patterns such as rhomboid pattern, Sassanian roundels pattern, geometric pattern, turtle back pattern, Fangsheng pattern（方胜纹）and so on.

There are only a few geometrical patterns preserved in the early Tang Dynasty, among which the most important is the rhomboid pattern. Rhomboid pattern often appeared in edging decoration, and compound with little flower, consistently show the shape of a whole and two halves, commonly decorated on the junctions of four slopes, niche edges, caisson trimmings, etc. Heart-linked pattern inherited from West Asia which mostly appears on the niche edge in the first-phase caves of the Tang Dynasty in Mogao Grottoes, but they have undergone a certain "localized" transformation — the heart-shaped pattern has been transformed into three-split small petals and became the petal pattern. In addition, lotus petals, peony petals, tendrils and other combinations of flowers, directly arranged into long edges, was also a unique decorative pattern in the Tang Dynasty. These petals are arranged in a whole and two halves in lengthways, or split in half-petal distributed on left and right of the long decorative areas.

There are also some decorative patterns with unique functions, such as flame pattern, which been being used to show the Buddha backlight, head light; The hanging triangle pattern, which was used to express the caisson four fringes, symbolizing the hanging curtain of the canopy.

On the whole, although there are many subcategories of this part of patterns, but the volume in the decoration system of the early Tang Dynasty is not as large as the previous two types. This paper focuses on the analysis of the dominant linear patterns and round patterns of the Tang Dynasty, so the other patterns are not analyzed in detail.

3. Linear Pattern

Linear decorative patterns in Mogao Grottoes mostly appeared in all kinds of trimmings. Because most of the caves in the Tang Dynasty were truncated pyramidal ceiling and have a niche in the west wall, so these trimming ornaments were mostly concentrated on the four sides of the caisson, the four slopes, the four walls, the inner and outer edges of the west niche, the backlight and the head light of the statues, and some of them appeared in the trimming ornaments on statues. These trimmings, according to the different decoration area, can be roughly divided into linear pattern and curved pattern. Among them, the curved patterns are mostly concentrated on the headlight and backlight edges, which are decorated in annular and semi-annular areas.

The linear patterns in the early Tang Dynasty were mainly scrolling grass pattern, and the main theme was pomegranate. Grapes in the early Tang Dynasty continued the form of the Sui Dynasty, but only occasionally painted, and did not become popular in later time.

3.1 The frame structure

The linear patterns in the early Tang Dynasty were mostly distributed on the long trimming parts, and the pattern framework was wavy and two in a group repeated. However, it is just a general term. Since the linear pattern appeared at the beginning of the Northern Dynasty, then to the early Tang Dynasty, the frame kept in a mild wavy form (Fig. 4). However, in the high Tang Dynasty, a large number of nearly circular linear patterns appeared, such as Cave 217, 172 and 444. Instead of the early slender and sparse style, people tried to make the curl of flowers and branches and leaves neat, and tried to maintain the consistency between the pattern units. The regular and rigor framework were associated with influences from Central Asia and India.

In the early Tang Dynasty, the frame of the scrolling grass pattern was mostly single-line frame, and the frame was simply wavy. All the flowers and leaves grow on the frame, but they do not overlap with the frame, instead they completely avoid the frame curve, so that the frame curve can be presented very clearly. In the high Tang Dynasty, the small branches of the scrolling grass pattern would interweave with the frame, and a large number of compound scrolling grass patterns with multiple vine frame appeared. This way was adopted in edge decoration in many caves (Fig. 5). Its source of such mutiple frame can be traced back to the Warring States period linear decorative patterns. In the Tang Dynasty, it became richer and thicker.

Curvilinear scrolling grass pattern often appeared in the head light during the Tang Dynasty, which were

Fig. 4 Pomegranate scrolling grass pattern on lintel in Cave 220 of Mogao Grottoes in early Tang Dynasty

Fig. 5 Cave 444 of Mogao Grottoes in the high of Tang Dynasty decorated with pomegranate scrolling grass pattern on niche edge

arranged around the head light center, and the flowers and leaves were distributed in the gaps on this basis. For example, in the head light edge of the west niche in cave 208, 217, 41 and 444, most of them were pomegranate scrolling grass (Fig. 6). The unit shape of the scrolling grass pattern is distributed in the halo-like area of the head to remind people of the head light pattern in the Gupta period of India. However, the scrolling grass pattern in the Gupta style is basically distributed in a static style, while the scrolling grass pattern in Dunhuang at this time is connected by wavy lines, forming a flowing trend.

If we look at the traditional Chinese linear modeling, in the late Neolithic Age, we can see a large number of scrolls, water waves and other line modeling decoration in the decorative patterns on pottery. This kind of wavy line has a very strong adaptability when forming decorative effect, filling decorative surface and related areas, and the shape is relatively casual and has a fluid vitality. A large number of lacquer ware unearthed in the Qin and Han Dynasties, especially in the Han Dynasty, were basically decorated with long lines, forming different images such as clouds and animals, all of which sought to be smooth and flexible (Fig. 7). As shown in Fig. 8, on the unearthed square pot dated to the Western Han Dynasty, there are smooth cloud air patterns. Clusters of lines are used to express the flow of cloud air, and the shape is smooth and graceful. Such patterns

Fig. 6 Head light in Cave 217 of Mogao Grottoes during the high Tang Dynasty

Fig. 7 Hubei Yunmeng Shuihudi unearthed Qin Dynasty lacquer plate

Fig. 8 Lacquered pottery Fang of the Western Han Dynasty

have a profound influence on the curves of later generations, especially on the vitality of lines. In the Southern Dynasty of China, a large number of pictures on bricks used long wavy lines to create images of clouds, clothing, dragons, and plants and so on. From the Han Dynasty to the Six Dynasties, we can see that these stretching and flowing clouds or other related patterns often constitute a dynamic space. Because of this, the patterns did not fill the painting area, but left a lot of gaps, forming a specific spatial layout, showing the ethereal feelings. This form of composition became a tradition for quite a long time. It can be seen from the decorative patterns of Mogao Grottoes in the early Tang Dynasty that the sparse lines style, which began in the Western Wei Dynasty, is added more details, such as the integration of a variety of plant themes, forming the unique characteristic of scrolling grass pattern in the early Tang Dynasty. These scrolling grass patterns soon became full, leaving less and less space in the decorative area. The relationship between patterns and backgrounds was very different from that in the early period. In the middle and late Tang Dynasty, the pattern continued the full trend of the high Tang Dynasty, and there was almost no blank area in the pattern filling. From the early Tang Dynasty to the high Tang Dynasty, we can see that the ethereal and elegant decorative style of the Han Dynasty was rapidly developing to the complex and full trend. The lively dynamic style formed by the linear patterns still maintained.

3.2 Partial modeling feature cases

3.2.1 The evolution of honeysuckle pattern

The honeysuckle pattern mostly appeared in the Northern Dynasty, and the traces of honeysuckle could also be seen in the lines of the early Tang Dynasty. For example, the caisson trimmings in Cave 329 (Fig. 9) are end-to-end scrolling grass patterns. The frame of the main body can be seen with wavy branches. Although the curling branches and leaves obviously have rich characteristics of the scrolling grass pattern from the Tang Dynasty, but the leaves on the main vine still have the appearance of honeysuckle pattern from the Northern Dynasty. Except in the Tang Dynasty there were a lot of changes on modelling, such as the leaves became more thick, edges became smooth and rolled, and honeysuckle leaves developed to the direction into tendrils and embedded peony petals, it integrated into bud shape, which seems to be the proof that honeysuckle pattern changed into scrolling grass in the Tang Dynasty. The attempts made in the scrolling pattern here show how the pattern evolves in the process of development, which some combinations developed from germination while others disappeared.

3.2.2 Flower and grass pattern and small flower pattern trimming

The edge decoration of flower and grass pattern is a special kind among many linear patterns, which was popular in caves during the high Tang Dynasty. It is different from the common style of two in a group repeated scrolling grass pattern, because there is no particularly clear curve frame, and it is difficult to define its pattern cycle unit. And even though the whole pattern is generally two in a group repeated pattern, but each unit is substantially different — it is difficult to find a complete and exact cycle unit, either the type or the shape. At present, there are about 15 high Tang Dynasty caves preserved flower and grass pattern which are visible (Fig. 10). These trimmings are distributed on the junctions of the four slopes, as well as niche edges, headlight edges, caisson edges and other parts (Fig. 11). The flowers and leaves of each unit are as different as possible, making the overall flowers look more diverse and varied. If the pomegranate scrolling grass pattern used certain model makes each cycle unit same, while the flower and grass pattern showed the artists' creativity: in a large area of pattern, left and right sides are symmetrical, full of changes, any two partials are not the same, painters

Fig. 9 The caisson in Cave 329 of Mogao Grottoes in the early Tang Dynasty

Fig. 10 Flower and grass pattern in Cave 66 of Mogao Grottoes of the high Tang Dynasty

tried their best to display the creativity and diversity. This kind of pattern from a certain extent reached the zenith of complication of linear patterns.

Decorative pattern in early Tang Dynasty and high Tang Dynasty, taking caisson for example, there is no any two caves are completely same, and the painters even did not use the same model that looks relatively similar. Flower and grass pattern centered massive different elements in small area, through this we can see the whole trend — flower and grass pattern represents in a certain sense the expression of high Tang Dynasty caves decoration and aesthetic ideal, this means high Tang Dynasty grottoes decoration contains extremely rich creativity and the pursuit to the aesthetic spirit of innovation. This is a kind of epitome of the early Tang Dynasty aesthetic ideal.

4. Round pattern

Flower medallion pattern was one of the most important patterns in Mogao grottoes decoration on wall and colored sculpture in the early Tang Dynasty. This is consistent with the widespread popularity of flower medallion pattern in society at that time. In the early Tang Dynasty, the pattern of flower medallion was dominated by the four-petaled cross shaped structure and the eight-petaled "米" shaped structure. The basic structure of the flower medallion pattern in the early Tang Dynasty consists with four main petals, which are decorated based on this cross structure. From the persimmon pedestal pattern of the cross structure in the bronze mirror of the Han Dynasty, it has a certain origin relationship with the flower medallion pattern of the cross structure popular in the early Tang Dynasty in terms of shape (Fig. 12). In the early Tang Dynasty, both the preference for cross structure and the modeling characteristics of single lotus petals are related to each other. In the high Tang Dynasty, the pattern of flower medallion dominated by eight-petaled structure, and more petal structure appeared. The internal changes from the early Tang Dynasty to the high Tang Dynasty, the frame changed from cross structure to "米" shaped structure, and the level of petals developed from single layer to multi-layers.

Flower medallion pattern formed the shape of "medallion" in real sense in the early Tang Dynasty, which combined different plant patterns and themes to form a overlooking multi-layers flower, and decorated in the center of caisson well on the top of caves with the most gorgeous composition form. This core pattern leads the

Fig. 11 Flower and grass head light decoration in Mogao Grottoes, Cave 225 of the high Tang Dynasty

Fig. 12. Western Han Dynasty "China Daning" Ruishoubojuwen(瑞兽博局纹) pattern gilt bronze mirror

whole cave decoration. Buddha statue is the key element of cave construction and the visual core of worship, but in truncated pyramidal caves, for decoration the caisson center was paid the most intention — flower medallion and layers of fringes all concentrated here intensively, and the decoration on the junctions of four slopes and other parts in cave (in the characteristics of modelling, colour, etc), are followed by the caisson center artistic style.

4.1 Partial modeling features and main themes

In the early Tang Dynasty, the shape of the flower medallion still retained a small sense of traces of the late Sui Dynasty — that is, the shape characteristics of the simple overlooking lotus in the caisson: the four corners are a quarter lotus, and the whole design is almost without the difference of layers, all are double layer petals. Soon the lotus structure began to complicate and compound various themes. Of course, peony and tendrils had retained for a long time in the fusion of different. Peony with tendrils and side-curling lotus together form the most important identifying features of the Tang Dynasty flower medallion pattern, especially the early Tang Dynasty medallion flower pattern. At the same time, tendrils, peony petals, Ruyi (an S-shaped ornamental object, a symbol of good luck, in this case, the design more close to C-shaped) pattern, pomegranate, grape and other themes appeared (Fig. 13).

More and more peony, pomegranate, camellia and other flowers popular at that time appeared in the decoration patterns of Mogao Grottoes. Obviously, they do not have the symbolic meaning of Buddhist art, but are the expression of integration of religious art and secular life. Of course, the dominant position of lotus flower has not changed, but in addition to the central part of the head light of some Buddha statues and the lotus flowers under the feet of Buddhas and Bodhisattvas, lotus decorations in other places have no longer attached importance to the expression of religious significance, and are more inclined to pure aesthetic reproduction. Besides the lotus flower, a large number of decorative patterns popular in secular society were applied in the decoration of Tang Dynasty caves.

Compared with the early Tang Dynasty, most of structures inherited to the high Tang Dynasty. In the caves of the third period of the Tang Dynasty, the petals in many flower medallion patterns were complicated, that

Fig. 13 Cave 387 and Cave 205 of Mogao Grottoes in the early Tang Dynasty

is, various elements were decorated as rich as possible. However, in the caves of the fourth period of the Tang Dynasty, that is, the last period of the high Tang Dynasty, the artists were not satisfied with the enrichment of petals as "petals". Instead, they drew petals directly into small flowers and used small flowers to form large clusters of flowers. For the overall development of decoration, such a state is still a step to enrich the patterns as much as possible. This technique reached its peak in the fourth period of the Tang Dynasty, and no further complex forms appeared (Fig. 14, Fig. 15). In the Middle Tang Dynasty, single flower style was preserved, but the whole design tended to be simpler and clearer, such as in cave 201 which used camellia as component for flower medallion, we still can see such attempts to be continued and inherited, then in the late Tang Dynasty it became into wreath type flower medallion pattern, which means the flower medallion composed by six to eight

Fig. 14 Flower medallion pattern head light in Cave 444 of the high Tang Dynasty in Mogao Grottoes

Fig. 15 The caisson(sketch) in Cave 185 of Mogao Grottoes in the high of Tang Dynasty

little flowers. At that period the creativity was dropped broadly in the artistic sense, and the decoration pattern became more simpler and duller.

4.2 Half flower medallion

Half flower medallion pattern often did not appear alone, but in groups to form trimmings. A decorative area can be occupied alone by it on the vertical surfaces of caisson center. In edging decoration, a common pattern is "one whole and two halves", which is also called "a whole and two broken" by some scholars. It means that in such edging decoration, the larger complete flower medallions are arranged in the central area, and half flower medallions are placed in the gaps on both sides (Fig. 16). These circular units consists of whole flower medallions and half flower medallions are so called a whole and two halves; or it is composed all by half flower medallions, such as in cave 387, 335, 340, 341, 321, 320, 79 caisson edging decoration, etc. The half medallion edging decoration appeared in a large number of caves in the early Tang Dynasty, which could appear in any edging parts. It was more widely used and more representative than the scrolling grass pattern, especially in the clothes edging decoration, the half flower medallion pattern was often used instead of the scrolling grass pattern. Here the half flower medallion pattern already has the characteristic of linear pattern.

Fig. 16 The caisson edge decoration in Cave 217 of Mogao Grottoes in the high Tang Dynasty

Such linear decorative trimming can be extended into decorative area, and in the Tang Dynasty textiles like this kind of flower medallion pattern brocade was very common. Most of the brocade patterns have main flower medallion and subsidiary flowers. The main flower is full and has rich layers, it usually has eight petals based on "米" shaped structure; the subsidiary flowers were filled in the gaps, usually small and based on cross structure; its edges are usually in concave shaped, in order to adapt the main flower's full outline. By doing this, the spaces among them are shaped in harmonious smooth ring.

Half flower medallion pattern has some extreme examples; in some design the artist purposely make the half flower medallion pattern into an independent pattern, not a half, which means the half is a pattern. So this kind of pattern cannot be doubled into a whole flower medallion. This kind of cases can be seen in cave 41, 31, 79 caisson edging decoration (Fig. 17).

Fig. 17 The caisson edging decoration in Cave 31 of Mogao Grottoes in the high of Tang Dynasty

4.3 Rounded flower pattern

There is a kind of decoration pattern — its outer contour is round, but not completely developed from center into a radial symmetry pattern, the cross structure, six petals or eight petals can't be seen, and often has the side

Fig. 18 Head light in Cave 217 of Mogao Grottoes during the high Tang Dynasty

of flowers, even used half flower medallion pattern and scrolling grass pattern as edging decoration, etc. The outer contour is still round shape, the inside structure often used different designs to fill up (Fig. 18). Because the overall decorative area is round, so it is called as analogous flower medallion pattern. This kind of pattern usually appeared in cave murals like the head light, body light and other parts.

4.4 Sassanian roundels medallion pattern

Sassanian roundels pattern was a popular pattern from the Northern Dynasty to the Sui and Tang Dynasties. Originating from the Persian textile pattern, the basic composition of the pattern is composed of small beads like circles, often depicting animal or hunting patterns inside the rings. This kind of pattern is very common in archaeological discoveries of textiles from the Northern Dynasty (some may be introduced from Persia). It appeared in the middle of Sui dynasty murals in Mogao grottoes, in cave 292, 425, 402, 401, 390, 394, 56 all have Sassanian roundels pattern. Some of them have hunting or animal motifs like the original Persia style, at the same time some have flower patterns in the centers. For example, in cave 425 west niche edging Sassanian roundels pattern decoration, we can see clearly the cross structure, this decoration style continued in the first period of the Tang Dynasty caves. Because Sassanian roundels pattern is round, in this paper, it is classified as a round pattern.

Circled Sassanian roundels pattern was very common in the Tang Dynasty's craftwork. The center of the circled Sassanian roundels pattern were usually filled with hunting scene, double goats, double deer and so on or plant patterns. In Mogao Grottoes, a large number of Sassanian roundels pattern began to appear in the Sui Dynasty and continued in the early Tang Dynasty. After that, the circular Sassanian roundels pattern was rarely seen, instead more grottoes used linear Sassanian roundels pattern. This kind of linear beads pattern was used as border decoration on the junctions of various surfaces in caves, such as the four slopes' junctions and west niche edge decoration, or to enrich the decorative surfaces with multi-layer linear patterns, such as caisson edge decoration. A large number of Sassanian roundels pattern appeared in the edge decoration during the Sui Dynasty caves, which in some sense assisted the round honeysuckle pattern, together they pushed the change of the overall development trend of decorative patterns from linear to round, that is, the division of units was more clear. However, the Sassanian roundels pattern with hunting scene and winged animals as themes which originated from western region were gradually replaced by plant patterns, which present a more peaceful and stable visual effect in the edge decoration of caves.

5. Color and technique

The color composition of decorative patterns is consistent with the overall color layout in the cave. In the murals of the Northern Dynasty, the Western Regions style was popular, paying attention to the three-dimensional expression of the figures, and the layers of the decorative patterns were also rich. Usually, honeysuckle pattern was depicted by the method of color-gradation technique, which seems thick and rich. The Northern Zhou and

Sui dynasty murals were given more attention in the line-drawing effect and manifested more plane decoration, such as in cave 292, 420, 207 and so on. The honeysuckle pattern lines were white or black, the internal filling color seems to be dispensable, the pattern has fluent and elegant visual effect, this effect mostly attributed to the line arrangement. In the period of the early Tang Dynasty, because of the influence of Central Plains new style, represented by cave 220 murals, craftsman gave more and more attention to lines which the colors and lines were combined closely, and the colors also show the ups and downs of the structure change, cohesion and shading subtlety. These exquisite details were what the previous dynasties painters could not achieve. The typical early Tang Dynasty colors are: malachite green, azurite, and ochre, light blue, dark green, earth red. Some caves in the early Tang Dynasty retained a large area of earth red and ultramarine combination from the Sui Dynasty, but after the further development in the Tang Dynasty, the main color of the caves was mainly clear and gorgeous cyan color, and the decorative patterns were mostly azurite and malachite green, supplemented by earth red, white and other elements. Some of the colors became brown and black due to oxidation and discoloration, creating the effect we see today. But through these changes, it is still possible to feel the gorgeous, refreshing atmosphere of the decorations in the early Tang Dynasty. The external environment of Mogao Grottoes is desert with little vegetation coverage. The ancient pilgrims would have been shocked by the turquoise they saw and the gorgeous beauty of the Buddhist world when they entered the grottoes after passing through the sandy landscape. From this point of view, the artists in the early Tang Dynasty were very good at creating a unique religious atmosphere.

In the high Tang Dynasty grottoes, the overall color characteristics were further strengthened on the basis of the early Tang Dynasty. The combination of azurite, malachite green and red also continues to dominate in grottoes, but the red color such as earth red and cinnabar had increased significantly compared with the previous caves. The decoration patterns compared with the early Tang Dynasty became more complex, azurite, malachite green, green, red, earth red and white and gold together forming a magnificent, graceful scene. Because of the oxidation and discoloration, there are a lot of black and brown in the mural, which are what we see today. The high Tang Dynasty caves, especially the fourth period, are more subtle in their lines and colors than most of the early Tang Dynasty caves. Take Cave 172 as an example. Many details in the murals can be viewed from a distance or appreciated closely — the artists took great efforts to the overall effect and details. As a result, it has almost infinite amount of details and worthy parts for worshipper or viewers to appreciate slowly, if he or she stays in the cave for a long time.

Summary

Since Mogao Grottoes started from the Northern Dynasty, the decorative patterns showed an important feature dominated by line modeling. More than ten different types of linear honeysuckle patterns constituted the main body of the cave decoration in the Northern Dynasty, while honeysuckle patterns themselves had smooth line modeling characteristics. The Sui Dynasty was a period that connects the past to the future. The long linear honeysuckle is a kind of unit decomposition of honeysuckle pattern, and it is also the inheritance of the linear pattern. The early Tang Dynasty inherited the form of linear patterns from the Northern Dynasty to the Sui Dynasty, but there were great changes in the details. In particular, the early form dominated by honeysuckle was replaced by a variety of curly and richly layered scrolling grass patterns, which brought novel vitality. In the early Tang Dynasty pattern, the main structure of many scrolling grass patterns or flower medallion patterns are relatively sparse, and the modeling elements emphasized on the smooth momentum of "line" very much. No matter what plant theme are, the artists always gave priority to line modelling, the linear structure decided the pattern's density and details; colors complement with lines, which enhanced the diversity. Looking back to the Qin and Han Dynasties, we can see the traditional Chinese decoration and painting art, the line modeling was very important. Through the curves and changes of lines, a kind of vibrant atmosphere is created, which so-called "vivid spirit" level. Such a tradition has continued to influence the later generations of artistic creation. Artists in the Tang Dynasty embedded aesthetic spirit in lines on decorative pattern, it has independent life energy which the Western Region patterns' lines do not have, to pursue the spiritual expression. The branches and leaves of the decoration pattern also conform to the era of local aesthetic characteristics, which is one of the Tang Dynasty painters' great creativity manifestations.

Since the Sui Dynasty, round honeysuckle pattern have appeared in many edging decoration — that is, honeysuckle leaves rotated or combined based on a cross structure into round honeysuckle pattern to decorate the long edges. The transformation of honeysuckle pattern from pure linear pattern to unitized independent round pattern can be regarded as an attempt in the process of transformation from linear pattern to round pattern. At the same time, honeysuckle pattern itself was decomposed and transformed to some extent, then integrated into the partial details of flower medallion pattern and scrolling grass pattern in the early Tang Dynasty. In addition, the popularity of Sassanian roundels pattern from the Persian region had contributed to the shift from linear pattern to round pattern to some extent — Sassanian roundels pattern themselves have the characteristics of round pattern, and the closed boundary formed a circular area, which is often composed by two in a group repeated edging decoration. When a large number of such round pattern were prevalent on art crafts during the Tang Dynasty, they also showed the same tendency in the decorative patterns of Mogao Grottoes. In the early Tang Dynasty, the edge decoration containing the round shaped elements coexisted with the scrolling grass pattern, at the mean time gradually occupied the dominant position. The caisson flower medallion pattern led the overall cave decoration and showed a strong trend of the round shaped pattern. Grotto's decoration and secular art crafts are inextricably linked. The significant improvement of the related craft level also had an important impact on the decorative forms in the murals. Dunhuang patterns are also the concrete expression of the social spirit that tended to flourish in the early Tang Dynasty. Round shape carries plump, rich, prosperous atmosphere, was a symbol of the Tang Dynasty. Round type decorative patterns had close relationship with textile, gold and silver vessels in the same period, and they have very similar shape characteristics, including multiple cultural birthplaces. The source of those elements includs Asia, India, central Asia, the Yangtze River and other places, where the farming cultures were highly developed and the societies and cultures were very prosperous. These exotic styles were continuously integrated with local artistic expression and aesthetic tendency in the early and high Tang dynasties.

The unification of the Tang Dynasty played an important basic role in the stable development of society and economy, in which culture and art tended to be prosperous and stable. Temple murals were very prosperous in the Tang Dynasty, and a large number of outstanding painters were engaged in Buddhist mural creation. For example, 33 painters mentioned in the Records of *Famous Paintings of Past Dynasties*（《历代名画记》）had carried out temple murals or related Buddhist painting creation. In addition, 29 painters were recorded in the *Youyangzazu-sitaji*（《酉阳杂俎・寺塔记》）, which shows that the main works of painters at that time were in temples. Temples were closely related to people's lives, which were important places for cultural activities. A large number of painters from Chang'an brought the developed skills and various artistic styles from the Central Plains to Dunhuang. The artistic creation of Central Plains, with the capital city as the center, had a strong radiation effect to the surrounding areas. This artistic style had spread to all society, and the creation of temple murals also shows unique creativity under the influence of such social atmosphere and ideology.

From the linear pattern to the round pattern, it is a process of transformation of decoration ideas, and also a process of evolution of artistic expression style, which means that the unit structure is becoming more and more complex. As a part of the Tang Dynasty art, the decorative art of Mogao Grottoes evolves from single to multiple, simple to complex, and sparse to flourishing in terms of theme, shape, pattern and color. The integration and application of flowers with various themes and the increasing complexity of flower elements all show the overall development trend of decorative patterns from different aspects, showing more and more richness and complexity. This is consistent with the trend of art, literature and even social life in the Tang Dynasty. Under this overall cultural environment, the decorative art in Mogao Grottoes manifested the trend to flourishing style.

隋唐染织工艺在敦煌服饰图案中的体现

杨建军

清华大学美术学院

服饰图案以独特的形态装扮着敦煌石窟彩塑和壁画中的佛陀、菩萨、弟子、天王、力士、罗汉等佛国人物，以及被称为供养人的出资造窟的功德主（窟主）及其家族成员和各类大众人物。它不仅使彩塑和壁画的人物形象更加美丽传神，也使整个石窟艺术更加完整精彩。织、染、印、绣、绘、拼补等不同染织工艺形成的不同图案特征反映在敦煌服饰图案上，千变万化，异彩纷呈。下面结合文献记载，参照传世或出土实物，对应隋唐时期的染织工艺种类，对敦煌服饰图案进行扼要分析。

六世纪末，隋朝结束分裂割据的局面，统一疆域，促进社会发展，起到了承上启下的作用。随后的唐代社会全面发展，进入中国古代最为辉煌的时代，人称大唐帝国。与其相一致的是这个时期中国染织工艺高度发达，随着制作技术的不断进步和显著提高，这时的染织艺术也进入了极盛时期。高超的染织工艺，在敦煌隋唐壁画和彩塑繁丽华美的服饰图案上得到了完整和客观的体现。

一、丝织工艺

染色工艺和织造技术的发展使隋唐时期丝织品的花纹变化多样，色彩丰富华丽。

流行于两汉及魏晋南北朝的传统平纹经锦，仍然是隋唐时期重要的丝织品种类之一。隋代第292窟南壁彩塑菩萨的半臂纹饰，于近似棋格状的方格内填饰联珠小团花纹，整齐而灵巧（图1），与其相类似的还有隋代第427窟中心柱南侧彩塑佛陀的内衣纹饰，它们表现的都是当时流行的精巧华美的棋格联珠小团花纹锦。对比收藏于日本东京国立博物馆的格子花平纹经锦（图2），不管是棋格形几何纹，还是格内填饰的联珠小团花纹乃至色彩，都与敦煌彩塑服饰中描绘的棋格联珠小团花纹极为相像。所以，敦煌隋代服饰图案中这类规矩的几何小团花纹，理应表现的是隋唐时期仍在生产的平纹经锦。1967年出土于新疆吐鲁番阿斯塔那唐代墓葬的小团花纹锦，是难得的这一时期的联珠小团花纹锦实物（图3）。这种通过西域传入欧亚各国的金银器图案，通过艳丽的色彩表现在织锦上，显得更为富贵华丽。这件唐锦表现

图1　敦煌莫高窟隋代第292窟南壁彩塑菩萨半臂纹饰

图2　格子花纹锦，七世纪，日本东京国立博物馆藏

图3　小团花纹锦，唐，中国新疆维吾尔自治区博物馆藏

的联珠环绕的多瓣小菊花形象，与隋代第427窟、第292窟棋格内的小团花如出一辙。据此，从日本正仓院收藏的传世丝织品和中国新疆博物馆收藏的出土实物两方面，都证明了敦煌壁画和彩塑绘制的服饰图案与当时丝织品的高度一致性。此外，从花纹、地色以及织物背面纹样的清晰程度分析，现存实物可能就是属于文献记载中的“蜀江锦”❶。那么，敦煌隋代彩塑上描绘的应该也是属于“蜀江锦”风格的平纹组织、经线起花的几何联珠小团花纹锦。

随着织造技术的改进，织物图案也不断创新。很多隋唐时期彩塑和壁画上的服饰图案，真实反映了制作技术革新带来的装饰风格的变化。大约于隋代之前已经出现的斜纹经锦和平纹纬锦，流行于隋唐时期。例如隋代第420窟西壁龛口南北两侧彩塑胁侍菩萨的裙饰图案，在浑厚古朴的色调上描绘金线、白线，表现了飞马奔腾、人兽交战的激烈场面和织物图案的细部特征，恰到好处地丰富了层次关系（图4）。从其特征看，可以断定它表现的正是属于斜纹经锦或平纹纬锦的隋代典型织物。图案以联珠纹组成圆环单位，内饰各种飞禽、走兽。这是盛行于波斯萨珊王朝时期的域外图案，传入中国后，在隋唐时期得到广泛流行和发展，形成了圆环状联珠纹内安置对鸟、对兽的新型图案。这种祥禽瑞兽置身于花环团窠的图案形式，很可能就是始创于初唐、兴盛于盛唐及中唐的“陵阳公样”❷。1972年在新疆吐鲁番阿斯塔那唐代墓葬出土的绛红色对飞马纹锦（图5），即是属于此种纹样的丝织物。隋代第420窟彩塑菩萨裙饰上描绘的联珠圈纹中填饰飞马纹、狩猎（驯兽）纹的服饰图案，反映的正是当时中外文化交流背景下产生的联珠飞马狩猎（驯兽）纹锦。另外，初唐第217窟西壁南侧大势至菩萨覆膊衣纹饰，其整齐斜向排列的棋格纹及单纯色彩，也清晰表明斜纹经锦的特征。

唐代中外交流频繁，不断开拓创新，反映在丝织图案上比隋代更加追求富于多变和华丽清新的气质。1968年在新疆吐鲁番阿斯塔那唐代墓葬出土了一件精美的红地花鸟纹锦，具有典型盛世唐锦的富丽华美特征：花团锦簇、禽鸟飞翔、祥云缭绕、情趣盎然。正如唐朝诗人王建在《织锦曲》中描写的“红缕葳蕤紫茸软，蝶飞参差花婉转。”❸的热闹景象。其生动的形象、活泼的布局、热烈的色彩，呈现出一

图4　敦煌莫高窟隋代第420窟西壁龛口南侧彩塑胁侍菩萨裙饰

图5　绛红色对飞马纹锦，唐，中国新疆维吾尔自治区博物馆藏

---

❶ 指中国汉代以后四川地区出产的名锦。

❷ 唐初年间，陵阳公窦师伦创瑞锦、宫绫，章彩奇丽，蜀人谓之“陵阳公样”。

❸ 中国社会科学院文学研究所古代文学室.《唐诗选注 上、下册》[M]. 北京：北京出版社，1978：441.

派富贵吉祥的祥和气氛，代表了唐代斜纹经锦的高度水平（图6）。将其对照隋代第427窟前室南壁西侧彩塑天王甲裳图案和中唐第159窟西壁龛内南侧彩塑菩萨长裙纹饰（图7），无论是生动严谨的造型、动静和谐的构图、绚丽华美的色彩，还是自由热烈的气氛，三者都极为相似。由此可知，隋代第427窟的彩塑天王甲裳图案和中唐第159窟的彩塑菩萨长裙纹饰，与隋唐时期极为高超的丝织技艺高度相符。

图6　花鸟纹锦，唐，中国新疆维吾尔自治区博物馆藏

图7　敦煌莫高窟中唐第159窟西壁龛内南侧彩塑菩萨长裙纹饰

隋唐时期的织造技术不断变化和发展，初唐时发明了斜纹纬锦，摆脱了以往织造小花纹的限制，可以自由织造大花纹乃至整幅单独图案，从而使表现繁复纹饰和丰富色彩成为现实。日本正仓院收藏着一件传世的八世纪斜纹纬锦琵琶袋，是极为典型精彩的唐风美锦。与此锦水平相当的唐锦，不少被表现在敦煌隋唐服饰图案上。例如，隋唐时期仍然流行的汉地传统团花图案，随着斜纹组织、纬线起花织造技术的不断提高和成熟，其形态也不断变化。它在传统花形基础上，持续丰富创造，先后融合本土牡丹、莲花以及域外海石榴等特点，逐步形成多样、富丽、饱满的堪称唐代图案经典的宝相花[1]，这种变化结果客观反映在了敦煌服饰图案中。例如中唐第159窟西壁龛内北侧彩塑菩萨裙饰上描绘的大团花图案，正是这种唐代极为流行的寓意“宝相庄严”的宝相花纹（图8）。其丰腴的宝相花和绚美的色彩，与日本正仓院收藏的缥地大唐花纹锦琵琶袋背面局部纹饰极为相似（图9），表现的也是当时非常名贵的丝织品。以上精美的丝织实物及与之特征相近的敦煌服饰图案，充分显示出唐代斜纹纬锦达到了极高水平。同时，斜纹纬锦织造技术的日臻完善，也使广泛流行于唐代的与宝相花齐名的卷草[2]成为重要的服饰图案新形式。图10是日本正仓院收藏的缥地大唐花纹锦琵琶袋正面局部纹饰，其曲折委婉、翻转自如、枝叶缠绵的蔓草型装饰，正是风靡当时的唐代卷草纹。由此看来，盛唐第194窟西壁龛内南侧彩塑天王铠甲卷草图案（图11）和同窟西壁龛内北侧彩塑菩萨裙饰卷草图案，以及中唐第159窟西壁龛内南侧彩塑菩萨帔帛卷草图案和晚唐第149窟西壁龛内南侧彩塑天王甲裳卷草图案等，表现的都是类似缥地大唐花纹

[1] 唐朝代表性装饰纹样，取其“宝相庄严”之意。它起源于东汉，以后与佛教意义结合而程式化。造型上多将莲花、菊花、牡丹等自然形态进行艺术加工，使之成为富有装饰性的理想花纹。构成上多以四向或多向对称放射的形式组成圆形、方形、菱形、多边形等团花纹。因其流行于唐代，也称“唐花”。

[2] 受来自西亚波斯金银器纹饰的影响，由隋以前忍冬纹发展而来的蔓草型图案。因其盛行于唐代，故又称“唐草”。

图8　敦煌莫高窟中唐第159窟西壁龛内北侧彩塑菩萨长裙纹饰

图9　缥地大唐花纹锦琵琶袋背面（局部），八世纪，日本正仓院藏

图10　缥地大唐花纹锦琵琶袋正面（局部），八世纪，日本正仓院藏

图11　敦煌莫高窟盛唐第194窟西壁龛内南侧彩塑天王铠甲卷草图案

锦琵琶袋这样的宝相花（唐花）卷草（唐草）纹锦等著名的唐代丝织绝品，极其繁复绚丽，代表了唐代斜纹纬锦的最高成就。

随着金银工艺的发展和对富贵时尚的追求，锦中加金的织金技术在盛唐时期广为流传，形成织金唐锦的绚丽华贵之风，这种风尚也出现在敦煌唐代服饰图案中。初唐第328窟西壁龛内富丽堂皇的彩塑半跏菩萨裙饰（图12），表现的就是这种锦中加金的织金及印金、绣金等工艺，极尽奢华。大约于晚唐时期，斜纹纬锦丝织技术又有了新的变化，从唐式纬锦[1]过渡到辽式纬锦[2]。敦煌晚唐之后的服饰图案，本

---

❶ 指出现于初唐，流行于盛唐、中唐直至晚唐的斜纹纬锦。

❷ 指出现于中唐，流行于晚唐乃至五代的半明经型斜纹纬锦。因此类织锦大量出自辽代，故称为辽式纬锦。

该不乏表现辽式纬锦的实例。只是晚唐特别是五代之后敦煌石窟日趋衰落，昔日风光渐逝。同时，仅从绘制的图案效果看，也难以区分同为斜纹纬锦的两者在基本组织结构上的区别。

唐朝的贵族妇女非常喜爱轻薄柔美的纱、縠、罗类丝织服饰，这种风尚也形象地表现在敦煌服饰图案上。例如初唐第321窟东壁菩萨裙饰，从其飘逸的衣纹、轻柔的质感、透明的材质等特点分析，表现的是唐代名贵的纱、縠织物。其上典雅清新的团状小散花纹，正是隋唐时期普遍流行的大众化印花图案。晚唐第14窟南壁的菩萨裙饰，是唐代极为流行的施以彩印或彩绣十字形散花的透明纱、罗裙的真实写照（图13）。诗人王建在《宫词一百首》中写道："缣罗不着索轻容，对面教人染退红。衫子成来一遍出，明朝半片在园中。"[1]对照敦煌壁画可以想见，薄如蝉翼的纱服、罗衣，衬托婀娜娇柔的体态，愈加妩媚动人。

图12　敦煌莫高窟初唐第328窟西壁龛内北侧彩塑半跏菩萨裙饰

图13　敦煌莫高窟晚唐第14窟南壁菩萨裙饰

二、刺绣工艺

"日暮堂前花蕊娇，争拈小笔上床描。绣成安向春园里，引得黄莺下柳条。"[2]活灵活现、生动传神，这是唐朝诗人胡令能在《观郑州崔郎中诸妓绣样》中赞美绣品的诗句。可见，同丝织工艺一样，隋唐时期的刺绣工艺也有了极大进步。

隋唐刺绣在针法上突破了战国及秦汉以来辫子股以及短针联结的切针等传统技法的局限，出现了接针、劈针、齐针和平金绣等新针法，以往复经纬的直线纹绣为特征的平绣工艺，在唐代也很快发展起来。同时，表现色彩深浅变化的分层退晕的套针或戗针，也被广泛运用在刺绣艺术中。这在大量出土及传世的绣品中得到了充分证实，不胜枚举。

敦煌隋唐时期的服饰图案也有大量表现刺绣工艺的，如服装的衣领、裙边等部位，以及条帛、帔巾等饰物，经常施以绣花。因而，衣领或裙边及帔帛、帔巾等是敦煌服饰图案表现刺绣工艺的重点。隋代第62窟西壁龛顶北侧持拂天女的帔巾，应是一件华丽的刺绣品。从描绘效果分析，纹样轮廓应该是用传统辫子股针法或由此发展而来的劈针（接针的一种）绣制，然后在部分轮廓内选用同类色线进行填绣，素雅而清新。这件帔巾是先分别绣制透明的纱、罗类绣片，然后再相互拼缝而成（图14）。

初唐时非常流行在薄如蝉翼的服装边缘进行刺绣，这一风尚被如实地定格在敦煌壁画中。例如初唐第321窟东壁北侧胁侍菩萨的透明天衣，搭肩绕臂，飘逸柔美。条状边饰内是以二方连续形式顺向秩序排列的半花型图案。花形规整，色调素雅。图案均采用退晕手法，色彩过渡均匀，色阶清晰明确，具

---

[1] 彭定球，等.《全唐诗》第302卷影印本［M］. 上海：古籍出版社，1986：3446。

[2] 彭定球，等.《全唐诗》第727卷影印本［M］. 上海：古籍出版社，1986：8325。

有明确的刺绣工艺的戗针或平套针特点。再如初唐第334窟西壁龛内菩萨天衣（图15），也展现了清秀飘逸之美。菩萨以双手将轻柔的天衣捧在胸前，两端顺势飘在身侧。红地衬托蓝、绿花纹，对比强烈。边缘清晰秀美的纹饰，大概是通过劈针、齐针等不同针法刺绣而成。盛唐时期，绣品同织锦一样，也追求华美富丽。盛唐第45窟西壁龛内南侧彩塑阿难尊者交领偏衫边饰（图16），表现的是唐朝典型的带状波浪式卷草图案，花叶顺势翻转，动感非常强烈。现实中这类领部边饰图案，通常以刺绣手法制作。从其饱满典雅的绘制效果看，应是选用蓝、绿、白等协调色线，以接针或劈针以及齐针、套针等混合平绣的方法，于纱、罗类底料上绣制完成。盛唐第328窟西壁龛内北侧彩塑迦叶尊者内裙边饰，色彩热烈，金碧辉煌，应该仿自盛唐加金技术展现的华贵气氛。其中，以半宝相花型连续排列构成的红地边饰，在金色映衬下，极尽富丽之美。花形外轮廓施盘金绣，其内以戗针或套针满绣，应该是此裙边饰的制作工艺原型。进入中晚唐时期，刺绣工艺表现力更加丰富。晚唐第196窟中心佛坛上北侧彩塑菩萨的披帛（图17），自左肩至右肋斜向披饰。土红地色上的几何型石绿云雷纹，清爽素雅，精巧细致。从其间距较大（如果采用丝织工艺会造成浮线过长）的排列方式分析，理应表现的是刺绣工艺。同一尊彩塑菩萨的裙饰（图18），非常繁复华丽，内裙底边带状“一整二破”式团花纹，以及外裙自下而

图14　敦煌莫高窟隋代第62窟西壁龛顶北侧持拂天女帔巾

图15　敦煌莫高窟初唐第334窟西壁龛内菩萨天衣

图16　敦煌莫高窟盛唐第45窟西壁龛内南侧彩塑阿难尊者交领偏衫边饰

图17　敦煌莫高窟晚唐第196窟中心佛坛上北侧彩塑菩萨披帛纹饰

图18　敦煌莫高窟晚唐第196窟中心佛坛上北侧彩塑菩萨裙饰

上依次排列的带状波浪式卷草纹、云纹和散点式花叶纹，极有可能是依据当时精美的刺绣裙绘制而成。

不过仅从唐代彩塑和壁画上绘制的图案效果看，很难区分其表现的是绣品还是织物。因为多色彩锦也常作为衣领或袖口的边饰以及制成披帛、帛带等，所以除了根据少部分图案可以断定分属不同的制作工艺外（如以流畅线条表现物象的，基本可以确定为绣品），大多数图案效果同时适用绣、织两种工艺，难以明确划分。例如图15的菩萨天衣边饰中规律连续的半花型图案，以及明确显示出交错布列的联珠四瓣花和四出忍冬纹，如果将其长边半花相对、反复相连，可以形成完整的四方连续图案，与中国新疆博物馆收藏的小团花纹唐锦图案极为相似。其单纯色彩和连续后形成的纹样直排特点，也明确表现出唐初平纹经锦特征。就此分析，它也极可能是将属于“蜀江锦”风格的联珠小四瓣团花纹锦，裁为条状而制作成服饰边缘的。

三、染印绘工艺

随着染色技术的不断提高，隋唐时期的印染工艺在前代基础上有空前的发展。除了传统的蜡缬、绞缬、凸纹木版印花、碱剂防染印花等工艺外，夹缬、镂空纸版印花等新工艺于此时迅速流行开来。

木版印花在隋唐时期很普遍，分为单色印花和多色印花，单色较为多见。晚唐第85窟窟顶东坡楞伽经变中尸毗王本生故事人物的商人袍服图案，其简朴、古拙的装饰风格，很明确显现凸纹木版单色印花的特点（图19）。盛唐第199窟西龛内弟子袈裟纹饰为五瓣小团花（图20），红绿相间，素朴清雅。其相同花形以各自独立的布局、圆润敦厚的造型以及简洁的色彩，显示出木版多色直接印花的特点。不过，将其对比收藏于日本正仓院的传世白地花鸟纹夹缬绝，以及出土于敦煌藏经洞、现藏于英国维多利亚阿伯特博物馆的簇六团花夹缬绢（图21），可以看出，其质朴浑厚、简洁明快的装饰风格，又与唐代夹缬实物很相符。所以，它又极有可能表现的是唐代非常盛行的多彩夹缬工艺。夹缬、木版印花都是雕刻木板制作夹花版和印花版，图案风格容易出现相近特征。所以，单从绘制效果来看，难以将其明确定位表现的是夹缬还是木版印花。

绞缬在隋唐时期高度发达，达到极盛。李贺在《蝴蝶飞》中有“杨花扑帐春云热，龟甲屏风醉眼缬”[1]的诗句，这里的“缬”即绞缬。将盛唐第194窟西壁龛内彩塑菩萨腰间垂带（图22），与收藏于日本

图19　敦煌莫高窟晚唐第85窟窟顶东坡商人袍服

图20　敦煌莫高窟盛唐第199窟西龛内弟子袈裟纹饰

图21　簇六团花夹缬绢，晚唐—五代，英国维多利亚阿伯特博物馆藏

---

[1] 彭定球，等.《全唐诗》第392卷影印本［M］. 上海：古籍出版社，1986：4419。

图22　敦煌莫高窟盛唐第194窟西壁龛内彩塑菩萨腰间垂带

图23　黄地七宝纹绞缬绢，八世纪，日本正仓院藏

图24　敦煌晚唐莫高窟第9窟西壁天王长身皮甲

正仓院的八世纪黄地七宝纹绞缬绢（图23）相比照，这条垂带无疑表现的正是唐代高超的绞缬工艺。晚唐第9窟西壁天王着长身皮甲（图24），仿自吐蕃铠甲装束。当时吐蕃游牧民族地区，已经盛行以捆绞手法染制毛织物，其典型图案为简洁明快的十字花，这在留存下来的西藏绞染实物上可以得到证实。图25是广东省博物馆收藏的一件西藏明代十字花绞染氆氇❶，是极具代表性的游牧民族绞染毛织物。其绛红地白色团圈图案，于团圈内染制的就是西藏传统的十字花纹。从绘制效果来看，晚唐第9窟西壁描绘的天王长身甲，表现的正是于皮质甲身边缘镶缝毛织物边饰的吐蕃皮甲，其边饰内的绞染花纹，正是西藏、蒙古等游牧民族代代相传的绞扎十字花。从花纹形态特征及轮廓边缘色彩渗化效果等方面进行比较，都与广东省博物馆收藏的这件绞染氆氇非常吻合。

图25　西藏十字花纹绞染氆氇，明代，广东省博物馆藏

印染绘等工艺还表现在袈裟上，佛及弟子、僧侣的袈裟以山水为纹，名曰山水衲，象征三山五岳和五湖四海，表现远离尘世、亲近自然的超凡脱俗寓意。山水纹袈裟被大量绘制在敦煌石窟中。隋代第244窟彩塑阿难尊者身着的山水衲，表现的应为麻布绘纹的彩绘工艺。到中唐和晚唐时期，山水衲虽然仍以隋以来的僧人常服为依据，但山水纹比隋代简练细腻，多以石绿、石青等色表现层次丰富的青绿山水，也多没有田相格限制，以四方连续结构整幅描绘（图26）。当然，从壁画上绘制的中、晚唐时期袈裟看，排列有序、色块均匀的山水纹，也表现出唐代极为流行的镂空纸版印花特点。同时，织物中的一种称为“伊卡特”❷的扎经或扎纬织造技术，正好也适合织造这样的图案轮廓随意自由而排列规律有序的

❶“氆氇”是藏语言译，为藏族等游牧民族手工生产的一种羊毛织品，一般用于衣服和坐垫等。

❷伊卡特为“Ikat”音译，来自印度尼西亚语“mengikat”一词，意为捆扎。指先把经线或纬线局部扎紧防染，再进行经纬织造的工艺或织物。

山水纹。中、晚唐时期的山水衲纹饰，究竟是属于哪一种制作工艺，还是选用不同工艺分别制作完成，或者是多种工艺综合运用，有待进一步研究考证。

四、拼补工艺

此时期拼补工艺除了用于袈裟外，也还用于制作间色裙。隋及初唐时期间色裙的裙腰上提，裙体加长，彩条变窄，很好地表现出女性婀娜多姿的修长体态，这在敦煌石窟中得到了如实表现（图27）。

图26　敦煌莫高窟中唐第159窟西壁龛内北侧彩塑弟子山水衲

图27　敦煌莫高窟初唐第381窟北壁女供养人像

敦煌彩塑或壁画中表现的服饰图案，使现实社会的织、染、印、绣等工艺尽显其中。不同制作工艺形成不同的图案风格：或轻柔细腻，或华美富丽，或刚健俊逸，或纤细婉约。当然，敦煌历代服饰图案不是制作工艺的解说图，而是赋予想象和创造的绘画艺术。因而，时有画僧或画工不拘细节的发挥之笔，甚至出现与当时工艺水平不符抑或难以划分工艺所属之处。但是，它毕竟来源于现实生活，以当时的生活着用服饰为依据或参照。所以，分析过程中通过彩塑或壁画描绘的服饰图案，尽量寻找最可能的制作工艺印迹，对照文字记载和出土实物，从而发现其绘制的服饰图案，在极大程度上真实反映了不同时期装饰风格的变迁及制作技术的改进，从而弥补了纺织品不易保存而传世甚少和出土实物不足的缺憾。

总之，作为佛教艺术的敦煌石窟，它不仅真实记录了中国公元四至十四世纪特别是隋唐盛世佛教的博兴，而且全面、完整、系统地反映了当时社会生活的原貌、中外文化交流的景象及其容纳吸收的整合创新。其中丰富多彩的服饰图案不但表现了当时染织艺术的真实风貌，同时也是体现中国古代织造、印染、刺绣等染织工艺技术的珍贵史料。

参考文献：

[1] 敦煌文物研究所．中国石窟·敦煌莫高窟：第二卷 [M]．北京：文物出版社，1984.

[2] 敦煌文物研究所．中国石窟·敦煌莫高窟：第三卷 [M]．北京：文物出版社，1987.

［3］敦煌文物研究所．中国石窟・敦煌莫高窟：第四卷［M］．北京：文物出版社，1987.

［4］常沙娜．中国敦煌历代服饰图案［M］．北京：中国轻工业出版社，2001.

［5］新疆维吾尔自治区博物馆出土文物展览工作组．丝绸之路：汉唐织物［M］．北京：文物出版社，1973.

［6］敦煌研究院．敦煌石窟全集・24・服饰画卷［M］．香港：商务印书馆（香港）有限公司，2005.

［7］赵丰．敦煌丝绸艺术全集：英藏卷［M］．上海：东华大学出版社，2007.

［8］松本包夫．正倉院裂と飛鳥天平の染織［M］．京都：紫紅社，1984.

插图名称及来源：

图1敦煌莫高窟隋代第292窟南壁彩塑菩萨半臂纹饰

（敦煌研究院．敦煌石窟全集・24・服饰画卷［M］．香港：商务印书馆（香港）有限公司，2005：79.

图2格子花纹锦，七世纪，日本东京国立博物馆藏

（松本包夫．正倉院裂と飛鳥天平の染織［M］．京都：紫紅社，1984：82.）

图3小团花纹锦，唐，中国新疆维吾尔自治区博物馆藏

（新疆维吾尔自治区博物馆出土文物展览工作组．丝绸之路：汉唐织物［M］．北京：文物出版社，1973：38.）

图4敦煌莫高窟隋代第420窟西壁龛口南侧彩塑胁侍菩萨裙饰

（敦煌文物研究所．敦煌莫高窟：第二卷［M］．北京：文物出版社，1984：62.）

图5绛红色对飞马纹锦，唐，中国新疆维吾尔自治区博物馆藏

（常沙娜．中国织绣服饰全集　第1卷　织染卷［M］．天津：天津人民美术出版社，2004：136.）

图6花鸟纹锦，唐，中国新疆维吾尔自治区博物馆藏

（新疆维吾尔自治区博物馆出土文物展览工作组：丝绸之路：汉唐织物［M］．北京：文物出版社，1973：44.）

图7敦煌莫高窟中唐第159窟西壁龛内南侧彩塑菩萨长裙纹饰

（敦煌文物研究所．敦煌莫高窟：第四卷［M］．北京：文物出版社，1987：77.）

图8敦煌莫高窟中唐第159窟西壁龛内北侧彩塑菩萨长裙纹饰

（敦煌文物研究所．敦煌莫高窟：第四卷［M］．北京：文物出版社，1987：78.）

图9缥地大唐花纹锦琵琶袋背面（局部），八世纪，日本正仓院藏

（松本包夫．正倉院裂と飛鳥天平の染織［M］．京都：紫紅社，1984：1.）

图10缥地大唐花纹锦琵琶袋正面（局部），八世纪，日本正仓院藏

（松本包夫．正倉院裂と飛鳥天平の染織［M］．京都：紫紅社，1984：4.）

图11敦煌莫高窟盛唐第194窟西壁龛内南侧彩塑天王铠甲卷草图案

（敦煌研究院．敦煌石窟全集・24・服饰画卷［M］．香港：商务印书馆（香港）有限公司，2005：144.）

图12敦煌莫高窟初唐第328窟西壁龛内北侧彩塑半跏菩萨裙饰

（敦煌文物研究所．敦煌莫高窟：第三卷［M］．北京：文物出版社，1987：115.）

图13敦煌莫高窟晚唐第14窟南壁菩萨裙饰

（敦煌研究院．敦煌石窟全集・24・服饰画卷［M］．香港：商务印书馆（香港）有限公司，2005：191.）

图14敦煌莫高窟隋代第62窟西壁龛顶北侧持拂天女帔巾

（敦煌文物研究所．敦煌莫高窟：第二卷［M］．北京：文物出版社，1984：128.）

图15敦煌莫高窟初唐第334窟西壁龛内菩萨天衣

（敦煌研究院．敦煌石窟全集·24·服饰画卷［M］．香港：商务印书馆（香港）有限公司，2005：141.）

图16敦煌莫高窟盛唐第45窟西壁龛内南侧彩塑阿难尊者交领偏衫边饰

（敦煌文物研究所．敦煌莫高窟：第三卷［M］．北京：文物出版社，1987：130.）

图17敦煌莫高窟晚唐第196窟中心佛坛上北侧彩塑菩萨披帛纹饰

（敦煌文物研究所．敦煌莫高窟：第四卷［M］．北京：文物出版社，1987：182.）

图18敦煌莫高窟晚唐第196窟中心佛坛上北侧彩塑菩萨裙饰

（敦煌文物研究所．敦煌莫高窟：第四卷［M］．北京：文物出版社，1987：182.）

图19敦煌莫高窟晚唐第85窟窟顶东坡商人袍服

（敦煌研究院．敦煌石窟全集·24·服饰画卷［M］．香港：商务印书馆（香港）有限公司，2005：166.）

图20敦煌莫高窟盛唐第199窟西龛内弟子袈裟纹饰

（敦煌研究院．敦煌石窟全集·24·服饰画卷［M］．香港：商务印书馆（香港）有限公司，2005：143.）

图21簇六团花夹缬绢，晚唐—五代，英国维多利亚阿伯特博物馆藏

（赵丰．敦煌丝绸艺术全集：英藏卷［M］．上海：东华大学出版社，2007：196.）

图22敦煌莫高窟盛唐第194窟西壁龛内彩塑菩萨腰间垂带

（敦煌文物研究所．敦煌莫高窟：第四卷［M］．北京：文物出版社，1987：42.）

图23黄地七宝纹绞缬绢，八世纪，日本正仓院藏

（松本包夫．正倉院裂と飛鳥天平の染織［M］．京都：紫紅社，1984：87.）

图24敦煌晚唐莫高窟第9窟西壁天王长身皮甲

（敦煌研究院．敦煌石窟全集·24·服饰画卷［M］．香港：商务印书馆（香港）有限公司，2005：162.）

图25西藏明代十字花纹绞染氆氇，明代，广东省博物馆藏

（常沙娜．中国织绣服饰全集　第1卷　织染卷［M］．天津：天津人民美术出版社，2004：369.）

图26敦煌莫高窟中唐第159窟西壁龛内北侧彩塑弟子山水衲

（敦煌文物研究所．敦煌莫高窟：第四卷［M］．北京：文物出版社，1987：78.）

图27敦煌莫高窟初唐第381窟北壁女供养人像

（敦煌研究院．敦煌石窟全集·24·服饰画卷［M］．香港：商务印书馆（香港）有限公司，2005：130.）

The Embodiment of Dyeing and Weaving Technology of Sui and Tang Dynasties in Dunhuang Clothing Patterns

Yang Jianjun

Academy of Arts & Design, Tsinghua University

The clothing patterns dress up the Buddhas, Bodhisattvas, disciples, devarājans, vīras, arhats and other Buddhist figures on sculptures and in murals, as well as those people so called Gongyangren who mainly sponsored the cave(cave owner) and their family members including people from all walks who are known as donors in Dunhuang Grottoes. They not only make the characteristic of sculptures and murals more beautiful and vivid, but also make the whole grotto art more complete and wonderful. Different patterns formed by weaving, dyeing, printing, embroidery, painting, patching and some other techniques reflected on the patterns of Dunhuang clothing, which are variable and colorful. The following is a brief analysis of Dunhuang clothing patterns on the bases of literature records and handed-down or unearthed objects, also corresponding to the dyeing and weaving techniques in Sui and Tang Dynasties.

At the end of the sixth century, the Sui Dynasty ended the situation of balkanization, unified the land and promoted social condition, which played as a transition period. Then the Tang Dynasty developed in an all-round way and entered the most glorious era in ancient China, known as the Tang Empire. In line with this period, China's dyeing and weaving technology were highly developed. With the continuous progress and significant breakthrough of production technology, the dyeing and weaving art at this time also entered the peak period. The superb dyeing and weaving technology has been fully and objectively reflected in the colorful clothing patterns of Dunhuang murals and sculptures.

1. Silk weaving technology

With the development of dyeing and weaving technology, the patterns of silk fabrics in Sui and Tang Dynasties were varied and colorful.

The traditional plain warp brocade, which was popular in the Han, Wei, Jin, Southern and Northern Dynasties, was still one of the important silk fabrics in Sui and Tang Dynasties. The sculpture Bodhisattvas' elbow sleeve at the south wall of cave 292 dated to the Sui Dynasty is neatly and dexterously divided by a checkerboard like square pattern, decorated by Sassanian roundels with small round flower pattern in the center (Fig. 1). Similar to the decorative pattern on Buddha's underwear at the south side of the central pillar in cave 427, Sui Dynasty, all of them show the exquisite and gorgeous checkerboard Sassanian roundels with small round flower pattern which was popular at that time. Compared with the checkered plain warp brocade collected in the

National Museum of Tokyo, Japan (Fig. 2), whether it is a plaid geometric pattern, or added small round flower with roundels and colors in the squares, they are all very similar to the patterns on Dunhuang painted sculptures costumes. Therefore, this kind of regular geometric pattern with small flowers on costume in the Sui Dynasty should indicate that the plain warp brocade was still in production in Sui and Tang Dynasties. The small round flower brocade unearthed in a Tang Dynasty tomb at Astana, Turpan, Xinjiang, in 1967 is a rare discovery(Fig. 3). This beautiful piece of Tang brocade shows the pattern of multi petal chrysanthemum surrounded by roundels is similar to the pattern in the cave 427 and cave 292 dated to the Sui Dynasty. According to this, from the two proofs of the silk fabrics handed down from ancient times and collected in Shosoin of Japan and the unearthed objects collected in Xinjiang Museum of China, it is safe to say that the clothing patterns on Dunhuang murals and sculptures are highly consistent with the silk fabrics at that time. In addition, by analyzing the pattern, ground color and the arrangement on the back of the fabric, the existing objects may belong to the "Shujiang brocade"[1] recorded in the literature. Then, the painted sculptures in Sui Dynasty at Dunhuang should also be the Sichuan brocade style of plain weave, which used vertical thread to weave motif, checkered with roundels and small round flower in the center.

Fig.1 The pattern on Bodhisattva's elbow sleeve, south wall of cave 292, Sui Dynasty, Mogao Grottoes, Dunhuang

Fig. 2 Plaid brocade, 7th century, National Museum of Tokyo, Japan

Fig. 3 Brocade with flower pattern, Tang Dynasty, Xinjiang Uygur Autonomous Region Museum, China

With the improvement of weaving technology, the pattern of fabrics also constantly innovated. Many costume patterns on painted sculptures and murals from Sui and Tang Dynasties truly reflect the changes of decoration style brought by technological innovation. Probably before the Sui Dynasty twill warp brocade and plain weft brocade had already appeared, then became popular in Sui and Tang Dynasties. For example, the pattern on Bodhisattvas' dress at the north and south side of the west niche in cave 420 dated to the Sui Dynasty used gold and white lines to depict the scenes of flying horses and hunting scenes, as well as the detailed features of fabric patterns, which appropriately enriches the gradation of colour(Fig. 4). According to its characteristics, it can be certain that it is the typical fabric of Sui Dynasty which belongs to twill warp brocade or plain weft brocade. The pattern is made up of a circular units of beads, and various birds and animals are placed in the center. This pattern was popular in the Sassanian period of Persia. After it was introduced into China, it was widely accepted and developed in the Sui and Tang Dynasties and transformed into a new pattern which has

[1] It refers to the famous brocade produced in Sichuan after the Han Dynasty.

double birds or double beasts in Sassanian roundels. This could be the so called Lingyanggongyang[1] started at the early Tang Dynasty, popular in high Tang and middle Tang Dynasty. In 1972, the crimson double flying horses brocade unearthed from the Tang Dynasty tomb in Astana, Turpan, Xinjiang (Fig. 5) is the silk fabric with this pattern. The patterns of flying horses and hunting scenes(maybe animal taming) patterns depicted on the sculpture of Bodhisattvas' skirt in cave 420 dated to Sui Dynasty reflected the historical background of cultural exchange between China and foreign countries at that time. In addition, on the southern side of the west wall of cave 217 dated to the early Tang Dynasty, the pattern on the Mahasthamaprapta Bodhisattva's coat also clearly shows the characteristics of the twill warp brocade, oblique plaid patterns with solid color.

The frequent communication between China and foreign countries in the Tang Dynasty and the continuous development and innovation of technology reflected in the fact that the silk patterns were more changeable and gorgeous than those in the Sui Dynasty. In 1968, an exquisite flowers and birds pattern with red background brocade was unearthed in a Tang Dynasty tomb in Astana, Turpan, Xinjiang. It has the typical rich and gorgeous characteristics of Tang brocade in the prosperous age: flowers, flying birds, auspicious clouds and full of happy mood. As the Tang Dynasty poet Wang Jian described in *Brocade Melody* (《织锦曲》), "the red threads are luxuriant, the purple velvet are soft, the butterfly flies intermittently, and the flowers are graceful"[2]. It's a lively scene, this vivid image, lively layout and warm colors show a kind of auspicious atmosphere of wealth, which represents the high level of twill warp brocade in Tang Dynasty (Fig. 6). Compared with the pattern of painted Heavenly King's armor and clothes at the west side of the south wall in the front chamber of cave 427 dated to Sui Dynasty and the pattern of painted sculpture Bodhisattva's long skirt at the west niche of cave 159

Fig. 4 The ornaments on Bodhisattvas'skirt, south side of the west niche, cave 420, Sui Dynasty, Dunhuang Mogao Grottoes

Fig. 5 Crimson brocade with double opposite flying horses pattern, Tang Dynasty, Xinjiang Uygur Autonomous Region Museum, China

[1] In the early Tang Dynasty, gongdou of Lingyang created Ruijin, Gongling, ZhangCai Qili, which was called Lingyang Gongyang by Shu people.

[2] Department of ancient literature, Institute of literature, Chinese Academy of Social Sciences, anthologies of Tang Poetry (Volume I and II), Beijing: Beijing Publishing House, 1978, p. 441.

Fig. 6 Flowers and birds pattern brocade, Tang Dynasty, Xinjiang Uygur Autonomous Region Museum, China

Fig.7 Painted Bodhisattva's long skirt, the south side of the Western niche, cave 159, the middle Tang Dynasty, Mogao Grottoes, Dunhuang

dated to the middle Tang Dynasty (Fig. 7), these three are very similar in terms of vivid and rigorous modeling, dynamic and static composition, gorgeous colors, free and warm atmosphere. From this, we can say that the patterns of the colored sculpture Heavenly King's armor and clothes in cave 427 dated to the Sui Dynasty and the long skirt patterns of the colored sculpture Bodhisattva in cave 159 dated to the Mid Tang Dynasty are highly consistent with the extremely good silk weaving skills in the Sui and Tang Dynasties.

The weaving technology in Sui and Tang Dynasties was constantly changing and developing. In the early Tang Dynasty, when twill weft brocade was invented, it got rid of the limitation of weaving small patterns in the past, and could weave large patterns freely and even cover the whole brocade, thus making the performance of complicated patterns and rich colors became possible. A Pipa bag with twill weft brocade from the eighth century is collected in Shosoin, Japan. It is a typical and wonderful brocade of Tang Dynasty. Many of the Tang brocade of the same level are shown in Dunhuang clothing patterns of Sui and Tang Dynasties. For example, the traditional Chinese round flower pattern, which was popular in Sui and Tang Dynasties, changed along with the development of twill weave and weft pattern weaving technology. On the basis of traditional flower forms, it continued to enrich and change, successively integrated the characteristics of local peony, lotus and foreign sea pomegranate, and gradually formed a variety, rich and full pattern, Baoxiang[1] flower, which can be called the classic pattern of Tang Dynasty. This change is objectively reflected in Dunhuang clothing patterns. For example, the pattern of large round flowers depicted on the colored sculpture Bodhisattva's skirt in the north side of the west niche of cave 159 dated to the middle Tang Dynasty exactly manifested this kind of precious image pattern

[1] The representative decorative pattern of Tang Dynasty takes the meaning of solemn treasure（宝相庄严）. It originated in the Eastern Han Dynasty, and later combined with Buddhist art and stylized. In terms of modeling, lotus, chrysanthemum, peony and other natural forms are artistically processed to make them ideal decorative patterns. It is composed of round, square, rhombic and polygonal patterns in the form of four or multi-directional symmetrical radiation. Because of its popularity in the Tang Dynasty, it is also known as Tang flower.

which was very popular in the Tang Dynasty, it means solemn and precious (Fig. 8).The plump flowers and gorgeous colors are very similar to the ornaments on the back of the pipa bag collected in Shosoin, Japan (Fig. 9). It proves that this skirt was kind of precious and expensive silk clothes at that time. The Pipa bag and Dunhuang clothing patterns with similar characteristics fully showed that the quality of twill weft brocade of Tang Dynasty had reached a very high level. At the same time, with the improvement of weaving technology of twill weft brocade, scrolling grass pattern[1] also became available and popular on clothing decoration in Tang Dynasty. It has become an important new form of clothing patterns. Fig. 10 shows the partial ornamentation on the front of the Pipa bag with Tang flower pattern with light blue background collected in Shosoin, Japan. The curled vine pattern decoration with twists and turns, turning freely and lingering branches and leaves, is exactly the popular scrolling grass pattern in the Tang Dynasty. From this point of view, the pattern on guardian's armor(Fig. 11), south side of the west niche, and the pattern on Bodhisattva's skirt on the north side in the west niche of cave 194 dated to the middle Tang Dynasty, the pattern on Bodhisattva's silk shawl, south side of the west niche in cave 159 dated to the middle Tang Dynasty and pattern on guardian's armor, the south side of the west niche in cave 149 dated to the late Tang Dynasty are all similar to the Pipa bag, the Baoxiang pattern and scrolling grass pattern. They were famous silk fabrics in the Tang Dynasty, extremely complicated and gorgeous, representing the highest achievement of twill weft brocade in Tang Dynasty.

With the development of gold and silver processing technology and the pursuit of wealth and fashion,

Fig. 8 Painted sculpture Bodhisattva's long skirt on the north side of the west niche in cave 159 of the middle Tang Dynasty in Mogao Grottoes, Dunhuang

Fig. 9 The back side of the pipa bag with pattern of Tang Dynasty, collected in Shosoin, Japan, the 8th century

[1] Influenced by the patterns of silver and gold wares from west Asia and Persia, the curled vine pattern developed from honeysuckle pattern which appeared before Sui Dynasty. Because of its prevalence in the Tang Dynasty, it is also known as the Tang grass.

Fig. 10 The front side of Pipa bag, Tang flower pattern with blue background, 8th century, collected in Shosoin, Japan

Fig. 11 The curled vine pattern on painted sculpture guardian's armor on the south side of the west niche in cave 194, high Tang Dynasty in Mogao Grottoes, Dunhuang

the gold weaving technology which adding gold into brocade was widely spread in the high Tang Dynasty, forming a gorgeous and luxurious style of Tang brocade weaving, which also appeared in the Tang Dynasty clothing patterns in Dunhuang. In the west niche of cave 328 dated to the early Tang Dynasty, the magnificent colored sculpture Bodhisattva's skirt ornaments (Fig. 12) show the technique of gold weaving, gold printing and gold embroidery in brocade, which is extremely luxurious. Around the late Tang Dynasty, the weaving technology of twill weft brocade had a new change, from the Tang style weft brocade transition to Liao style weft brocade. The clothing patterns in Dunhuang after the late Tang Dynasty should have some examples of Liao style weft brocade. However, in the late Tang Dynasty, especially after the Five Dynasties, Dunhuang Grottoes gradually declined and the former prosperity gradually disappeared. At the same time, it is difficult to distinguish the basic structure between the two only from painting, they both are twill weft brocade.

The noble women of Tang Dynasty were very fond of silk gauze, Hu(wrinkled silk gauze) and Luo(net like thin silk gauze) clothes, which were also vividly shown in the patterns of Dunhuang costumes. For example, the Bodhisattva's skirt ornaments in cave 321 dated to the early Tang Dynasty, from its elegant patterns, soft texture, transparent material and other characteristics, shows the valuable fabric Sha and Hu in the Tang Dynasty. The elegant and neat round flower pattern was the popular printing pattern in Sui and Tang Dynasties. The Bodhisattva's skirt on the south wall of cave 14 dated to the late Tang Dynasty is a good example of the transparent Sha and Luo skirt at that time, which were very popular in the Tang Dynasty with color printing or color embroidery to have cross shaped scattered flowers (Fig. 13). The poet Wang Jian wrote in *One Hundred Poems for Court Lady*（《宫词一百首》）: "*the court lady do not want thick cloth like Jian and Luo, they ask for thin silk gauze, so thy teach people to dye the cloth and made clothes, it was so thin that next day just like half of them hanging in the garden.*" Compared with the Dunhuang murals, we can imagine that the Sha and Luo clothes were thin as cicadas' wings, set off the graceful and beautiful body, which is more charming and moving.

Fig. 12 The painted sculpture Bodhisattva's skirt on the north side of the west niche of cave 328 dated to the early Tang Dynasty in Mogao Grottoes, Dunhuang

Fig. 13 Bodhisattva's skirt, the south wall in cave 14 of Mogao Grottoes in Dunhuang

2. Embroidery

"In the twilight, the flowers in front of the hall are beautiful. A group of lovely embroidery women, eagerly doing sketching on embroidery bed. They embroidered a beautiful screen, quietly put it into the garden, make the warbler curious, fly away from the wicker for it[1]*."* Hu Lingneng, a poet in the Tang Dynasty, praises embroidery in his poem *Observing the Embroidery Women at Cui Lang Zhong's House in Zhengzhou*. It reveals that, like silk weaving, embroidery in Sui and Tang Dynasties also made great progress.

The embroidery of Sui and Tang Dynasties has broken through the limitations of traditional techniques such as braided strands and cut needles connected by short needles since the Warring States period and Qin and Han Dynasties. New needling techniques, such as needle joining, splitting, even gold embroidery developed rapidly in the Tang Dynasty. At the same time, needle linking was also widely used in embroidery art to show the change of color depth. This has been fully confirmed in a large number of unearthed and handed down embroidery works.

In the Dunhuang caves of Sui and Tang Dynasties, there are also a large number of embroidery patterns, such as on the collar, skirt trimming and other parts of the clothing, as well as scarf, sash and other accessories, often embroidered. As a result, the collar or skirt trimming, scarf and sash are the key points of Dunhuang clothing patterns to study this aspect. The scarf on the north side of the top of the west niche in cave 62 dated to the Sui Dynasty is a gorgeous embroidery. By the analysis of the painting, the pattern outline should be embroidered with traditional braid stitch method or split needle (a kind of connecting needle) developed from the traditional braid stitch method, and then select the same color thread to embroider in part of the contour, which is elegant and fresh. This piece of silk was first embroidered with transparent Sha and Luo embroidery pieces, and then stitched together with each other (Fig. 14).

In the early Tang Dynasty, embroidery on the edge of clothing was very popular, which was faithfully shown in Dunhuang murals. For example, the transparent clothing on the north side of the east wall of cave 321 dated to the early Tang Dynasty is elegant and beautiful. Inside the stripe trim is a half flower pattern arranged in a two-way sequence. The flower shape is regular and the color is simple and elegant. The color transition is continuous, and the color scale is clear and neat. It shows the characteristics of the opposite needle or flat

---

[1] Peng Dingqiu. Complete Tang poetry, Vol. 727, Shanghai: Ancient Books Publishing House, 1986, p. 8325.

needle with clear embroidery technology. Another example is the clothing of the painted Bodhisattva in the west niche of cave 334 dated to the early Tang Dynasty (Fig. 15), which also shows the soft and neat beauty. With both hands, the Bodhisattva held the soft heavenly garment in front of his chest, and both ends floated on his sides. The red ground sets off the blue and green patterns, and the contrast is strong. The clear and beautiful decorative patterns are probably made by different needling methods, such as splitting needle and aligning needle. In the high Tang Dynasty, embroidery, like brocade, also pursued its splendor. In the west niche of cave 45 dated to the high Tang Dynasty, the painted sculpture Ananda the cross collar edge ornament (Fig. 16), which shows the typical pattern of scrolling grass in the Tang Dynasty, and the flowers and leaves flip in accordance with the trend, and the dynamic is very strong. In reality, this kind of collar trimming pattern is usually made by embroidery. From the perspective of its full and elegant drawing effect, it should be finished on the Sha and Luo base materials by selecting the coordinated color lines of blue, green and white, and using the method of needle joining or splitting, as well as the method of flat embroidery such as parallel stitch and cover stitch. In the west niche of cave 328 dated to the high Tang Dynasty, the inner skirt trimming of the colored sculpture Kasyapa is decorated with warm and brilliant colors, which should be imitated from the luxurious atmosphere displayed by the technique of adding gold in clothing in the high Tang Dynasty. Among them, the red ground edge ornaments, which are arranged continuously in the form of semi Baoxiang flowers, are extremely beautiful against the golden background. The outer outline of the flower is gold plate embroidery, and the inner part is fully embroidered with opposite needle or cover needle, which should be the prototype of this skirt decoration. In the middle and late Tang Dynasty, the expressive ability of embroidery was more abundant. On the north side of the central altar of cave 196 dated to the late Tang Dynasty, the painted sculpture Bodhisattva scarf (Fig. 17) is decorated obliquely from the left shoulder to the right rib. The geometric green cloud and thunder pattern on the earth red ground color are fresh, simple and elegant, exquisite and meticulous. According to the analysis of the large space arrangement (if silk weaving technology is adopted, the floating line will be too long), it could be embroidery technique. The skirt ornaments of the same colored statue Bodhisattva (Fig. 18) are very complicated and gorgeous. The bottom edge of the inner skirt is decorated with one whole and two broken pattern, and the

Fig. 14 The scarf on the goddess of heaven on the top of the west niche in cave 62 dated to the Sui Dynasty in Mogao Grottoes, Dunhuang

Fig. 15 Bodhisattva's scarf decoration in the west niche of cave 334 dated to early Tang Dynasty in Mogao Grottoes, Dunhuang

Fig. 16 Painted sculpture Ananda cross collar robes and accessories on the south side of the west niche in cave 45 dated to the high Tang Dynasty in Mogao Grottoes, Dunhuang

Fig. 17 Painted sculpture Bodhisattva strips patterns on the north side of the central Buddhist altar in cave 196 dated to the late Tang Dynasty in Mogao Grottoes, Dunhuang

Fig. 18 Painted sculpture Bodhisattva's skirt on the north side of the central altar of cave 196 dated to the late Tang Dynasty in Mogao Grottoes, Dunhuang

outer skirt is arranged from bottom to top in scrolling grass pattern, cloud pattern and scattered flowers pattern, which are probably based on the exquisite embroidery skirt at that time.

However, it is difficult to distinguish between embroidery and fabric only according to the pattern effect of painted sculptures and murals. Because multiple color brocade is also used as the edge decoration of collar or cuff, as well as made into silk scarf and silk belt, etc., except a small number of patterns(for example, if the images are represented by smooth lines, they can be almost certain as embroideries), it is difficult to clearly distinguish the pattern that they belong to what production processes only by painting. For example, the regular and continuous half flower pattern in the edge decoration of painted Bodhisattva's clothes in Figure 15, as well as the four petal lotus and four honeysuckle patterns clearly showing staggered arrangement, if the long side half flowers are connected relatively and repeatedly, a complete four side continuous pattern can be formed, which is very similar to the Tang brocade pattern with small group patterns collected in Xinjiang Museum of China. The simple color and the straight arrangement of patterns formed after the continuous formation also clearly show the characteristics of plain brocade in the early Tang Dynasty. But, it is possible that this could be the Shujiang brocade style of small four petals round flower brocade cut into strips and made into the edge of clothing.

3. Dyeing, printing and painting

With the continuous improvement of dyeing technology, the printing and dyeing technique of Sui and Tang Dynasties had an unprecedented development on the basis of previous generations. In addition to the traditional printing processes such as batik, tie-dye, embossed woodblock printing and alkali resist printing, new technologies such as nipping and hollowed paper printing became popular at this time.

Woodblock printing was very common in Sui and Tang Dynasties, which can be divided into monochrome printing and multi-color printing. In the *Laṅkāvatāra Sūtra* on the east slope of cave 85 dated to the late Tang Dynasty, the clothing patterns of the characters in story painting of Buddha's past life as king Sivi on the east slope have a simple and primitive decorative style, which clearly shows the characteristics of single color printing with convex wood block (Fig. 19). In the west niche of cave 199 dated to the high Tang Dynasty, the pattern of the disciple's kasaya has five petals small round flowers (Fig. 20), red and green, simple and elegant. The same pattern with their own independent layout, round and thick shape and simple color, shows the characteristics of wood block multi-color direct printing. However, by comparing the flower and bird patterns on the printed silk cloth collected in Shosoin, Japan, and the six clusters flowers of clip printing silk cloth unearthed in Dunhuang library cave and now in the British Victoria Abbott Museum (Fig. 21), we can see that their simple and solid, lucid and lively decoration style is very consistent with the clip printing in the Tang Dynasty. Therefore, it is very likely to show the very popular multi-color clip printing craft in the Tang Dynasty. Clip printing and woodblock printing are both made of carved wood, and the pattern style is easy to appear with similar characteristics. Therefore, from the perspective of drawing effect alone, it is difficult to clearly define whether the performance is clip printing or woodblock printing.

Fig. 19 Merchant's robes on the east slope of cave 85 dated to late Tang Dynasty in Mogao Grottoes, Dunhuang

Fig. 20 Kasaya patterns of disciples in the west niche of cave 199 dated to the high Tang Dynasty in Mogao Grottoes, Dunhuang

Fig. 21 Six clusters of flowers with clip printing silk, late Tang to Five Dynasties, collected by Albert Museum, Victoria, UK

In the Sui and Tang Dynasties, tie-dyeing was highly developed and reached its peak. Li He's *Flying Butterfly*（蝴蝶飞）has "The butterfly flutters gently against the indoor tent, which exudes the warm breath of spring. In front of the bed was a splendid tortoise shell screen with her beautiful colored clothes on it.（杨花扑帐春云热，龟甲屏风醉眼缬）"[1] The "缬" in this poem is tie-dyeing. Comparing the waist belt of painted sculpture Bodhisattva's in the west niche of cave 194 dated to the high Tang Dynasty (Fig. 22) with the seven treasure pattern tie-dyeing silk dated to the eighth century (Fig. 23) collected in Shosoin, Japan, it undoubtedly shows the superb craftsmanship of Tang Dynasty. The heavenly king in cave 9 dated to the late Tang Dynasty has a long body leather armor (Fig. 24), which is imitated from the Tibetan armor. At that time, in the Tibetan nomadic areas, it was popular to dye wool fabrics by means of binding and twisting. The typical pattern was simple and bright cross flowers, which can be confirmed by the preserved Tibetan stranding and dyeing objects.

---

[1] Peng Dingqiu (Qing) et al., Complete Tang Poetry, Vol. 392, Shanghai: Ancient Books Publishing House, 1986, P. 4419.

Fig. 22 Painted sculpture Bodhisattva pendants in the west niche of cave 194 dated to the high Tang Dynasty in Mogao Grottoes, Dunhuang

Fig. 23 The seven treasures pattern on the yellow ground silk, 8th century, collected in Shosoin, Japan

Fig. 24 Long body leather armor of the heavenly king on the west wall of cave 9 in Mogao Grottoes dated to the late Tang Dynasty

Fig. 25 is a cross dyed woolen fabric called Pulu[1] dated to the Ming Dynasty in Tibet collected by Guangdong Provincial Museum. It is a representative hank dyed wool fabric of nomadic nationality. Its crimson ground and white circle pattern is dyed in the traditional Tibetan cross pattern. From the perspective of drawing effect, the long body armor of the painted heavenly king depicted on the west wall of cave 9 dated to the late Tang Dynasty is exactly the Tibetan leather armor with woolen fabric edging on the edge of the leather armor body. The twist dyeing patterns in the edge decoration are the twisted cross flowers handed down from generation to generation by nomadic peoples such as Tibet and Mongolia. Compared with the pattern features and the color permeating effect of the contour edge, it is very consistent with the Pulu collected by Guangdong Provincial Museum.

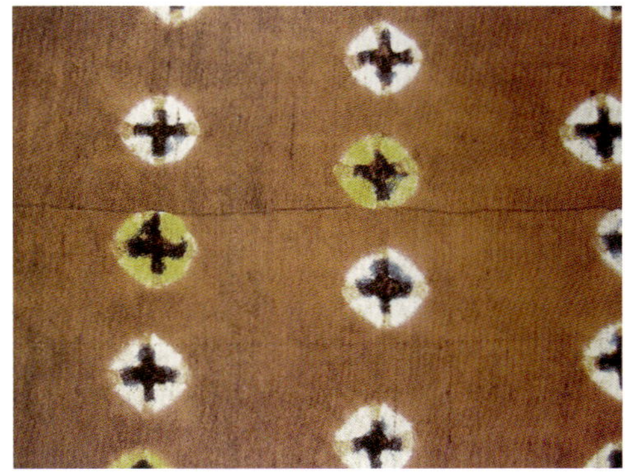

Fig. 25 Tibetan cross pattern Pulu, Ming Dynasty, collected by Guangdong Provincial Museum

Printing, dyeing and painting are also shown in the Kasaya. The Kasayas of Buddha, disciples and monks use mountain and water pattern to decorate, which are called Shanshuina(mountain water Kasaya), it symbolizes the Three Mountains and Five Mountains, and the Five Lakes and Four Seas. It shows the transcendent implication of being far away from the secular world and close to nature. There are many Kasayas in Dunhuang Grottoes painted with landscape pattern. In the Sui Dynasty, the painted statues Ananda's Shanshuina in cave 244 dated to the Sui Dynasty showed the painting technique on linen. In the middle and late Tang Dynasties, although the landscape pattern was still based on the common monk clothes since the Sui Dynasty, but were

[1] It is a kind of wool fabric translated from Tibetan language and produced by hand for Tibetan and other nomadic people. It is generally used for clothes and cushions.

more concise and delicate than those of the Sui Dynasty. Most of the green mountains and waters with rich levels were represented by the colors of azurite and malachite green, and most of them were depicted in a square continuous structure (Fig. 26). Of course, seen from the Kasaya from the middle and late Tang Dynasty on the murals, the landscape patterns with orderly arrangement and uniform color blocks also showed the characteristics of the hollowed out paper printing which was very popular in the Tang Dynasty. At the same time, one of the fabrics was called Ikat[1]. It is also suitable for weaving such landscape patterns with free pattern and regular arrangement. In the middle and late Tang Dynasty, it needs further research and study to identify which kind of manufacturing technology they used, or combined, or to use a variety of techniques comprehensively.

4. Patching technology

At this time period, the patchwork technique was used not only for Kasaya, but also for making striped color skirt. During the Sui and early Tang Dynasties, the skirt waist was lifted up, the skirt body was lengthened, and the color stripes were narrowed, which well showed the graceful and slender body shape of women. This was truthfully expressed in Dunhuang Grottoes (Fig. 27).

The costume patterns in Dunhuang painted sculptures and murals make the weaving, dyeing, printing, embroidery and other crafts in history be fully displayed. Different production techniques form different patterns: soft and delicate, gorgeous, vigorous and elegant, or delicate and graceful. Of course, Dunhuang clothing patterns

Fig. 26 Kasaya on a disciple of colored sculpture in the north side of the west niche in cave 159 dated to the middle Tang Dynasty in Mogao Grottoes, Dunhuang

Fig. 27 Female donors on the north wall of cave 381 dated to the early Tang Dynasty in Mogao Grottoes, Dunhuang

[1] Ikat is transliterated from the Indonesian word "mengikat", meaning binding. It refers to the process or fabric in which the warp or weft is partially tied up to prevent dyeing, and then the warp and weft are woven.

in the past dynasties are not illustrations of the production process, but painting art endowed with imagination and creation. As a result, there were painters or craftsmen who did not stick to the reality, and even appeared to be inconsistent with the level of technology at that time or it was difficult to classify which technique the craft belonged. However, they all originated from the real life after all and are based on the clothing dressed at that time. Therefore, in the process of analysis, we tried our best to find the most possible production process by analysing painted sculptures and murals carefully, comparing the written records and unearthed objects. We can find that the clothing patterns drawn by them truly reflects the changes of decorative styles and the improvement of production technology in different periods, thus making up for the defects that textiles are not easy to be preserved and rarely handed down and unearthed objects are insufficient.

In short, as a Buddhist art center, Dunhuang Grottoes not only truly recorded the prosperity of Buddhism from 4th-14th century A.D., especially in the prosperous period of Sui and Tang Dynasties, but also comprehensively, completely and systematically reflected the original appearance of social life, the scene of cultural exchange between China and foreign countries and the integration and innovation of accommodation and absorption. The colorful clothing patterns not only show the real style of dyeing and weaving art at that time, but also embody the precious historical materials of ancient Chinese weaving, printing and dyeing, embroidery and other clothing production technology.

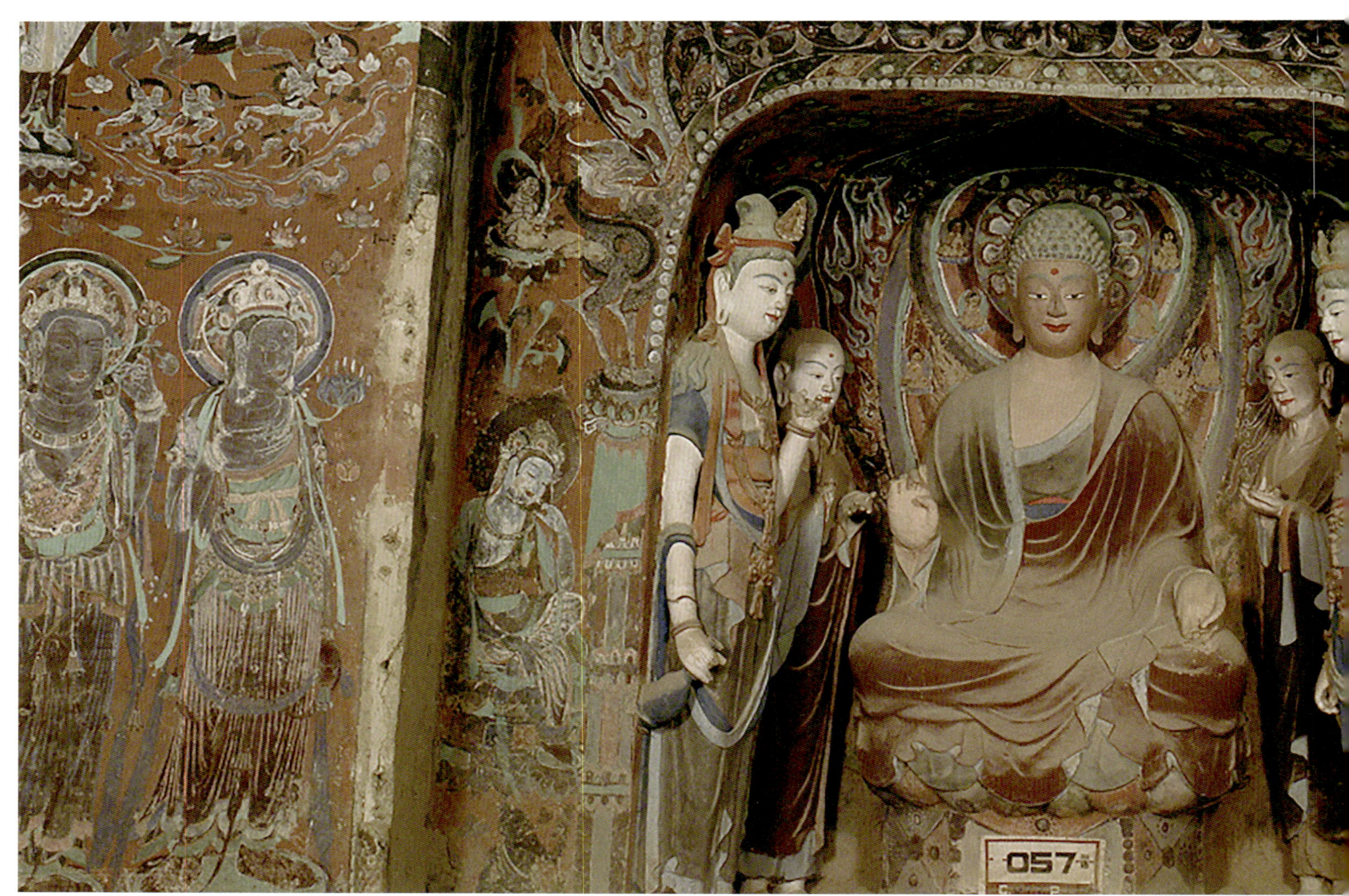
057

敦煌莫高窟初唐

第57窟

Dunhuang Mogao Grottoes Cave

57 of the Early Tang Dynasty

第57窟为覆斗顶方形中小型洞窟，初唐时期代表洞窟之一。洞窟顶部绘双龙莲花藻井，四周环绕飞天，四披绘千佛。西壁开双层圆券龛，沿袭了隋代洞窟的典型做法。龛内彩塑一佛、二弟子、四菩萨（失一），内层龛绘头光、背光、弟子和飞天，外层龛绘龛楣、龛柱、思维菩萨、飞天。龛外两侧各绘菩萨二身，上绘乘象入胎（北侧）和夜半逾城（南侧）等经典佛传故事情节，下绘供养菩萨十身。南北两壁中央和东壁门两侧各绘说法图一铺，四周和上部绘土红底千佛，下部为初唐（南壁）和晚唐（北壁）时期的供养人像（大部分已残毁）。甬道和前室壁画经晚唐重绘。

初唐第57窟的壁画在构图形式上沿袭了隋代的惯例，即为菩提树下一佛二菩萨二弟子，但是在绘画风格和人物表现技法方面开创了初唐时期的新气象。虽然此时还没有出现大铺的经变画，但是在说法图的人物构成、重叠关系、空间安排上更显复杂和多样。有的人物还带有隋代健硕的身形特征，但是线条处理和晕染层次方面更加柔和多变，画师常常通过身姿动态的曲线造型、手部动作的呼应变化和低眉顺目的顾盼眼神等细节刻画，增添人物的生动写实气息，让人感到亲切自然的真实感。特别是菩萨服饰中大量描绘的层层璎珞、飘带和披帛，延伸了人物的身体语言和气韵动向，使得人物的身体在直立挺拔中又拥有优美飘逸的风姿。

此窟服饰的描绘十分注重织物的细节表现，南北两壁、西壁龛外两侧、北壁、东壁服装款式绘制清晰，喜爱表现半透明的披帛和面料的效果。菩萨的上衣偏爱僧祇支的款式，下着长裙，腰部翻出，并喜爱于外翻的裙腰表现精美的纹样装饰，通身缠绕层层叠叠的披帛和璎珞，整体风格十分华丽。南北两壁、西壁龛外两侧、北壁菩萨服饰的织物纹样描绘非常精细，具体分布于僧祇支、裙腰、天衣上，这些纹样多为扎经染，并搭配联珠纹以及菱格纹，体现着唐朝纺织品对于波斯工艺和图案影响的吸收。

（文：崔岩）

Cave 57 is one of the middle sized caves with square main chamber and truncated pyramidal ceiling. It was built in the Early Tang Dynasty. The center of the ceiling has two painted dragons and a lotus flower, surrounded by flying Apsarases, the four slopes filled with Thousand Buddha motif. The west wall has a double-layer niche with round trimmings, which follows the typical practice in the Sui Dynasty. There are some painted sculptures in the niche: one Buddha, two disciples and four Bodhisattvas (one lost). The inner niche is painted with halo, mandorla, disciples and flying Apsarases, while the outer one is painted with lintels, columns, contemplate Bodhisattvas and flying Apsarases. Two Bodhisattvas are painted on both outer sides of the niche. On the upper part of the west wall has classic Buddhist stories which are the Great Conception (the north side) and the Great Departure (the south side) , beneath them are ten offering Bodhisattvas. On both sides of the east gate and the center of the north and south walls have one preaching scene painting respectively, around them filled with Thousand Buddha motif on red background. The lower area of the walls are images of donors dated to the early Tang Dynasty (south wall) and late Tang Dynasty (north wall) (most of them have faded or been damaged). The murals of the corridor and the front chamber were redrawn in the late Tang Dynasty.

The mural in cave 57 followed the custom of the Sui Dynasty in the form of composition, that is, one Buddha, two Bodhisattvas and two disciples under Bodhi tree. However, it created a new atmosphere in terms of painting style and character feature. Although there is no large scale Jingbian (Sutra illustration) painting during this period, the composition of characters, overlapping relationship and spatial arrangement became more complex and diverse. Some characters still have strong body features of the Sui Dynasty, but the line processing and coloring technique are more flexible and changeable. Painters often use dynamic curvilinear body posture, different hand movements, merciful and beautiful facial expression and other details to add the vividness and realistic feelings to the characters to make people feel a kind of natural sense reality. In particular, a large number of keyūra, ribbons and uttariya (披帛， similar to silk made scarf) depicted on Bodhisattvas which extended the figures' attractiveness and charm, making characters' upright body smooth and have a graceful and elegant demeanor.

In this cave the ancient craftsmen paid great attention to clothes' depiction in the aspect of fabrics. The clothes on the north and south walls, the outer two sides of the west niche, the north wall and the east wall are clearly drawn, and the translucent effect of uttariya and textile was preferred and expressed. Bodhisattva's upper clothes preferred the style of Sankaksika (僧衹支), long antariya (裹裙， similar to long skirt) on the lower body and the waist wrap turned out so the exquisite decoration pattern is presented. The whole body is wrapped with many silk scarves and keyūras, which makes the overall style very gorgeous. The fabric patterns of Bodhisattvas' clothes on the north and south walls, the outer sides of the west niche and the north wall are very detailed, which are specifically distributed on the Sankaksika, waist wrap and uttariya. These patterns are mostly tie-dye, combined with beads pattern and rhombus pattern, reflecting the influence of Persian crafts and patterns on Tang Dynasty textiles.

(Text by: Cui Yan)

Avalokitesvara Bodhisattva's clothes

on south wall of the main chamber in cave 57 dated to the early Tang Dynasty at Mogao Grottoes

观世音菩萨服饰

莫高窟初唐第57窟主室南壁

初唐第57窟主室南壁说法图为敦煌莫高窟的经典壁画，特别是说法图中的观世音菩萨形象生动优美，因此有人将此窟誉为“美人窟”，此处选取的就是这身美丽的观世音菩萨像。此像极具典型性，菩萨的面容与体态既气定神闲又婀娜多姿，体现修长的S形体态特征。菩萨面容秀美、禅悦，柳叶眉，鼻梁挺直，嘴角含笑，妩媚动人。头戴宝冠，微微颔首，颈身垂饰璎珞，戴臂钏和手镯，左手上举至颈肩，轻拈璎珞串珠，右手托在胸前，似在慈悲施予。

在服饰效果图的画面处理上，着意将上半身画得实一些，下半身画得虚一些，在视觉上有前倾之势。菩萨上身穿袒右僧祇支，束腰带，外裹裙腰和长裙，裙带束结玉环，飘垂至莲花座前，身上环绕轻薄透明的天衣。整体服饰造型上紧下松，错落有致，丝绸面料精良而考究，并有联珠纹、小团花纹和富有扎经染色织物特点的图案装饰其间。尤为凸显的是菩萨身上的饰物，均用沥粉堆金的工艺制成，富丽华美又生动立体。宝冠以火焰为装饰主题，中间为化佛，两侧辅以日月装饰。同时，镶嵌各种珠宝的颈饰、耳饰、臂饰、手镯等光彩夺目。另有点缀在身体多个部位的璎珞及各色飘带，衣饰与配饰珠联璧合，相映生辉，静中有动，动中有静，整体上给人以圆润秀雅、雍容华贵之美感。

（文：刘元凤）

The painting on the south wall of the main chamber of cave 57 dated to the early Tang Dynasty is a classic mural in Dunhuang Mogao Grottoes. In particular, the image of Avalokitesvara Bodhisattva is vivid and beautiful. Therefore, some people call this cave as Beauty Cave（美人窟）. This image is very typical, and the face and posture of this painted Bodhisattva is calm and graceful. The body is slender S-shaped. He has an exquisite face, a Zen like delight, willow eyebrows, a straight nose and smiling lips. He wears a crown, bows his head forward slightly, and is decorated with necklaces, armlets and bracelets. He holds his left hand up to left shoulder, gently touching the beads on the keyūra, and holds his right hand in front of his chest, as if giving mercifully.

In the image processing of clothes effect drawing, the upper body is drawn more solid, the lower body is simpler, which has a forward trend in vision. This Bodhisattva's upper body wears Sankaksika, tied with a waist belt, the lower body wrapped with waist wrap and a long antariya. The belt is tied into a jade ring（结玉环，a type of knot）and floating down to the lotus seat, at the same time the whole body is surrounded by a piece of light and transparent cloth. The whole dress is tight at the upper body and loose at lower part, designed with balance. The silk fabrics are exquisite and full of details, which have Sassanian roundels pattern, small round flower pattern and pattern with the characteristic of tie-dye fabrics. The ornaments on this Bodhisattva used Plaster-Squeezing and Gilding technique, looking gorgeous and vivid. The decoration theme of the crown is flame, with Buddha in the middle and sun and moon on both sides. At the same time, the neckwear, earrings, arm ornaments and bracelets inlaid with many kinds of precious stones are dazzling. In addition, the keyūras and ribbons in different parts of the body, along with the clothes and accessories match with each other perfectly, created the feeling of movement in stillness and stillness in movement. On the whole, this Bodhisattva is gentle and exquisite, gorgeous and noble.

（Text by: Liu Yuanfeng）

图：刘元风 Picture by: Liu Yuanfeng

The pattern on Avalokitesvara Bodhisattva's uttariya and waist wrap

on south wall of the main chamber in cave 57 dated to the early Tang Dynasty at Mogao Grottoes

观世音菩萨披帛图案 裙腰图案

莫高窟初唐第57窟主室南壁

第57窟主室南壁的观世音菩萨所着僧祇支的主体纹样是团花纹。僧祇支底色为浅土黄色，主花为联珠八瓣团花，宾花为四叶小朵花，花型轮廓均用白色线条勾勒。主花和宾花用白色线条相连形成散点式四方连续纹样，结构严谨，内容充实。缘边底色为深绿色，联珠八瓣团花纹左右相连，与主纹样相互呼应，丰富其艺术效果。

菩萨外裹裙腰底色为石青色，整体图案以菱形为骨架进行连缀排列，构成四方连续。菱形骨架用工整的排线方式表现，营造出视觉上的模糊感。而缘边使用深褐色为底色，以白线勾勒出联珠十二瓣小团花纹，用一破为二的形式交错排列，搭配几何纹样的裙身产生动静结合的装饰效果。

（文：王可）

The main pattern of Avalokitesvara Bodhisattva's Sankaksika on the south wall of cave 57 is round flower. The color is light yellow, the main flowers have eight petals and the minor flowers have four petals. They are all connected by white lines to form a scattered pattern with four in a group repeated. The structure is rigorous and the content is rich. The trimming background color is dark green, and the eight petal round flowers surrounded by beads pattern are placed in a line, which echoes with the main pattern, enriching its artistic effect.

Bodhisattva's waist wrap is dark blue, and the pattern is basically composed by rhombus, forming a four in a group repeated arrangement. The rhombic frame is displayed in neat lines, creating a sense of fuzzy vision. The trimming uses dark brown as the background color and white lines to form the 12-petal small round flowers which are surrounded by beads, staggered arrangement in the way of one broken into two, matching with the antariya geometric patterns and producing a dynamic and static decorative effect.

(Text by: Wang Ke)

图：王可 Picture by: Wang Ke

图：王可 Picture by: Wang Ke

供养菩萨服饰

莫高窟初唐第57窟西壁龛外北侧

The Offering Bodhisattva's clothes on the north outside of the west niche in cave 57 dated to the early Tang Dynasty at Mogao Grottoes

这里选取的是初唐第57窟西壁龛外北侧第一身供养菩萨。菩萨面相丰满，五官精致，眉间有白毫；双目微垂，凝神静气，并留以髭须，庄重典雅；身体修长，体态圆润，跣足立于莲台之上，站立的身体微侧向佛龛，似在听法。

服饰造型上，菩萨梳高髻，鬓角的发丝微微弯曲，头戴火焰纹宝冠。宝冠中央及两侧镶嵌绿色宝珠，两侧的宝珠下垂璎珞和红绿相间的流苏，并有带饰由两肩飘落。菩萨颈身有多层璎珞装饰，颈部的璎珞在肩部束结为宝珠，中央垂下流苏，并附有多条彩色飘带。层叠的链状璎珞在菩萨腹部交叉，环绕腰部和身体两侧，显得繁复华丽。菩萨戴莲花纹耳环，腕部戴手镯，左手上扬轻拈珠串，右手下垂轻握绕身飘落的天衣，仪态万方。上身穿袒右僧祇支，饰以菱格联珠纹，腰束金属革带，下穿长裙，裙带束结圆环，外裹的裙腰因扎紧而边缘形成有韵律的折叠波浪形，流畅的衣纹表现出面料悬垂的质感。整体服饰色彩配置以壁画现状的赭褐为基调，这是颜料氧化变色之后的效果，再点缀石青、石绿，轻重相宜，主次分明。

（文：刘元凤）

Here is the offering Bodhisattva on the north side of the west niche in cave 57 dated to the early Tang Dynasty. The Bodhisattva has a plump face, delicate facial features, and an urna between his eyebrows. His eyes are slightly close; mind is still, moustache beneath nose, looking solemn and elegant. The body is slender and mellow, standing on the lotus platform with bare feet; slightly face to the niche, as if listening to the Dharma.

In terms of dressing, the Bodhisattva has a high bun, the hairs on the temples are slightly curved, and the crown has flame pattern. The center and sides of the crown are inlaid with green pearls. On both sides of the crown, there are pendants and red and green tassels, ribbons dangling from both shoulders. The multi-layer keyūra is around the Bodhisattva's neck, it is connected on the shoulders by precious stones, and tassels hanging down in the center, along with many colorful ribbons. The keyūra chain meets at the Bodhisattva's belly, encircling the waist and both sides of the body, showing a complex and gorgeous appearance. The Bodhisattva wears lotus earrings and bracelets on his wrists. His left hand is up and gently touches beads, and his right hand is drooping holding the ribbon. The upper body wears Sankaksika, decorated with rhombus and beads pattern, and the waist is bound with metal leather belt. The lower body wears long antariya, the antariya belt tied as a circle knot. The waist wrap is tightly bound and forms a rhythmic folding wave shapes, the smooth folding shows the texture of the fabric. The overall color configuration of the clothes is the ochre brown following the mural status quo, which is the effect of pigment oxidation and discoloration, and then decorated with azurite and malachite green, with appropriate portion and good composition.

(Text by: Liu Yuanfeng)

图：刘元风 Picture by: Liu Yuanfeng

The Sankaksika pattern on offering Bodhisattva

on the north outside of the west niche in cave 57 dated to the early Tang Dynasty at Mogao Grottoes

供养菩萨僧祇支图案

莫高窟初唐第57窟西壁龛外北侧

菩萨上身穿袒右僧祇支，以浅土红为底色，主体图案为富于立体感的四方连续结构，用白色双道线条交叉形成的菱格纹构成基础骨架。右斜的条纹中间点缀着以白线勾边的褐色小圆点，显然受到了联珠纹的影响。菱格内填饰着规律的白色和褐色横条排线，中央用光影的手法表现立体凸起的圆点，显得别具一格。菩萨僧祇支缘边的底色为褐色，上有粗细相间的石绿色竖向条纹交错排列。整体图案色彩和谐典雅，结构明晰，疏密有致，充满节奏感和律动感。

（文：王可）

The Bodhisattva wears Sankaksika with right shoulder bared, and the background color is light earth red. The main pattern is a four in a group repeated structure rich in three-dimensional sense, and the rhomboid pattern formed by the crossing white double lines serve as the basic frame. The right-slanting stripes are interspersed with small brown dots with white lines, obviously influenced by the beads pattern. The rhomboid lattice is filled with regular white and brown horizontal lines, and the central part used light and shadow technique to show the three-dimensional raised dots, which is unique. The base color of the Sankaksika trimming is brown, and there are vertical malachite green stripes alternating thickness arranged. The overall pattern is harmonious and elegant in color, clear in structure and density, and full of rhythm.

(Text by: Wang Ke)

The pattern of Buddha's Sankaksika

on the south wall of the main chamber in cave 57 dated to the early Tang Dynasty at Mogao Grottoes

佛陀僧祇支图案

莫高窟初唐第57窟主室南壁

第57窟主室南壁中央绘弥勒说法图，位于中央的主尊弥勒佛结跏趺坐于狮子座上作说法相，因弥勒佛面部和手部的贴金被人为破坏，故模糊不清。佛陀外披袒右红色田相纹袈裟，内着浅石绿色僧祇支，腰间束带，这里选取僧祇支的缘边图案进行绘制整理。图案为半破式二方连续结构，以浅土红为底色，主体花型为浅石青色联珠团花纹和石绿色三瓣朵花纹，色彩清新淡雅，并以白色线条勾勒花朵轮廓和点缀花蕊。两种花纹分两列上下交错排列，使图案充满秩序感。

（文：王可）

The preaching scene of Maitreya is painted in the center of the south wall of the main chamber of cave 57. The main image Maitreya Buddha in the center sits on the lion seat with folded legs in preaching posture, the gilding on Maitreya's face and hands were damaged, not clear. The Buddha wears a red kasaya with field pattern and bares the right shoulder, light malachite green Sankaksika inside and a belt around his waist. Here I choose the trimming pattern of the Buddha's Sankaksika for drawing and analysis. The pattern is a half broken two in a group repeated structure, with light earth red as the background color, and the main body is light azurite round flower surrounded by beads pattern and malachite green three petals flower pattern. The color is fresh and clear, which used white lines to form the flower's shape and decorate the stamens. The two patterns are arranged alternately up and down in two columns to make the pattern full sense of order.

(Text by: Wang Ke)

图：王可　Picture by: Wang Ke

图：王可　Picture by: Wang Ke

帝释天服饰

莫高窟初唐第57窟西壁龛外南侧

Śakro devānām indrah' s clothes

on the south side of the west niche in cave 57 dated to the early Tang Dynasty at Mogao Grottoes

初唐第57窟主室西壁龛外南侧上部绘有佛传故事中夜半逾城的经典情节，表现的是释迦牟尼在成佛之前为悉达多太子时，生活极尽奢华，但长大成人后见到人间的生、老、病、死诸苦，经思索后决定出家寻求解脱之道。画面上太子乘马飞跃奔驰，马足下有四位天人各托一马足，四周祥云环绕，前有飞天翔舞奏乐，后有帝释天指引护持。据《佛本行集经・舍宫出家品》的记载，护世四大天王和帝释天得知太子出家，纷纷前来躬拜，其中“天主释提桓因，与其眷属一切诸天，百千万众前后导从，将天华鬘末香涂香，或复执持幡幢宝盖，或执种种诸妙璎珞，从彼三十三天而来”。

壁画中的帝释天面相俊美，表情安静平和，头戴宝冠，左手持麈尾，仰面前行，为潇洒飘逸之护法神形象。人物着中国传统经典的上衣下裳式服制，上为圆领衣套对襟大袖衫，领口和袖口皆有镶边装饰。内着圆领衣，其领口和窄袖与深棕色长裙色调一致，圆领衣的大身、对襟大袖衫的衬里与长裙的裙边皆用石绿，以浓淡变化穿插呼应，配色和谐丰富。下为齐踝长裙套裹围腰，在同期的佛造像与壁画中习见，体现出初唐时期外来服饰文化的影响与融合。整体服饰具有魏晋时期儒士装扮的风范，从侧面说明佛传故事的图像创作在流转过程中已经明显本土化的现象。

（文：崔岩、吴波）

On the south side of the west niche in the main chamber of cave 57 painted the classic story of Buddha's life the Great Departure dated to the early Tang Dynasty. It shows that when Sakyamuni was Prince Siddhartha before becoming a Buddha, he lived a very luxurious life. However, when he grew up, he saw the hardships of life, aging, illness and death. After thinking about them, he decided to become a practitioner and seek relief. In the painting, the prince is riding on a horse; under the horse's hoofs, there are four heavenly beings holding up horse's hoofs, surrounded by auspicious clouds. In front of the horse, there is a flying Apsaras dancing and playing music, and at the back of the horse Śakro devānām indrah giving guidance and protection. According to the records in the *Buddhist Scriptures of Buddha's Life*, the four great heavenly kings and Indrah came to worship when they learned that the prince decided to leave home. Among them, the scripture says "the Śakro devānām indrah, together with all the dependents, millions of celestial beings spread flowers, held banners, canopies, various wonderful jewels coming from thirty-three heavens."

In the mural, Śakro devānām indrah has a well-shaped face, a quiet and peaceful expression, a crown on his head, holding Zhuwei（麈尾， a kind of special fan）in his left hand, making him look like a graceful dharma protector. The clothes he wears is Chinese traditional. The upper body wears round collar shirt inside and Duijin (the two pieces making up the front of a Chinese jacket) coat outside with a pair of large sleeves, neckline and cuffs are all decorated with trimming. Under the coat he wears round collar shirt（圆领衣）, its collar and narrow sleeves are the same color to the dark brown long dhoti (the male skirt). The large body of the round collar shirt, the lining of the large sleeve coat and the dhoti trimming are all malachite green, the shade of color changes and the whole color is harmonious and rich. The lower part is the ankle length dhoti wrapped around the waist, which is commonly seen in Buddha sculptures and murals of the same period, reflecting the influence and integration of foreign clothes culture in the early Tang Dynasty. The whole dress has the demeanor of Confucians in the Wei and Jin Dynasties, which shows that the image creation of Buddhist stories has been obviously localized in the process of propagation.

（Text by: Cui Yan, Wu Bo）

图：吴波 Picture by: Wu Bo

敦煌莫高窟初唐

第68窟

Dunhuang Mogao Grottoes Cave

68 of the Early Tang Dynasty

第68窟为初唐时期创建的小型覆斗形窟，窟顶南、西、北披各存一部分千佛。西壁开平顶敞口龛，龛内彩塑一佛、二弟子、二菩萨，大部分经清代重绘，另浮塑释迦牟尼佛的头光和身光。龛顶绘一铺构图复杂、规模宏大的《妙法莲华经·见宝塔品》，龛内绘弟子像八身，两侧绘维摩诘经变，龛沿绘卷草纹边饰。龛外南侧绘普贤变，北侧绘文殊变。主室南北壁各绘经变一铺。

西龛内的整体色调为青绿色，搭配龛边缘卷草纹，整体色调十分清新。龛内壁面文殊菩萨与天女、大臣及众弟子服饰绘画用线流畅，服装形制清晰可辨，服饰纹样多为十字结构并满铺的小花。

（文：崔岩）

Cave 68 is a small-scale truncated pyramidal ceiling cave built in the early Tang Dynasty. There are Thousand Buddha motif on the south, west and north slopes. The west wall has a flat top open niche. Inside the niche, there are painted sculptures of one Buddha, two disciples and two Bodhisattvas; most parts of the sculptures had been repainted in the Qing Dynasty, the halo and mandorla of Sakyamuni Buddha are clay reliefs. On the top of the niche, there is a large-scale complex illustration of the *Lotus Sutra · the Emergence of Treasured Pagoda Chapter*. Inside the niche are eight painted figures of disciples, with *Vimalakirti Sutra* illustration on both sides, and the niche is decorated with scrolling grass pattern（卷草纹）. The south side of the niche is painted with Samantabhadra Bodhisattva tableau, and the north side is painted with Manjusri Bodhisattva tableau. The north and south walls of the main chamber are painted with a sutra illustration respectively.

The overall color in the west niche is turquoise, with scrolling grass pattern on the edge of the niche; the overall tone is very fresh. On the inner wall of the niche, the clothes of Manjusri Bodhisattva, Heavenly Lady, ministers and disciples are painted with smooth lines, and the shape of the clothes are clear and discernible. Most of the clothes patterns are cross structure and covered with florets.

（Text by: Cui Yan）

The pattern on the painted sculpture Ananda's kasaya
in the west niche of cave 68 dated to the early Tang Dynasty at Mogao Grottoes

彩塑阿难袈裟图案
莫高窟初唐第68窟西壁龛内

西壁龛内彩塑释迦牟尼佛的十大弟子之一阿难尊者，虽然面部为清代重绘，但服饰保留了初唐原妆，这里将其袈裟图案整理出来。阿难尊者内着右衽宽袖偏衫，外披田相纹袈裟，即称为僧伽梨（Saṅghāṭi）的大衣。其披着方式为袒右式，本应绕搭至左肩后的袈裟一角翻折后自然垂下，搭至左臂，但因系结的绚连缀肩后的纽所以形成三角形造型。画师按照佛教律典的规定和实际制作要求绘出袈裟形制，整体为土红色田相七条衣，缘边（四周边缘）与叶（中间的横条和竖条）有四瓣花纹，坛（田相格）内绘六瓣花纹，除了装饰之外，这些纹饰所处的位置说明其具有缝缀加固袈裟的作用。

（文：崔岩）

In the west niche stands a painted sculpture Ananda, one of the ten major disciples of Sakyamuni Buddha. Although his face was redrawn in the Qing Dynasty, but his clothes remains the original color of the early Tang Dynasty. Here, the pattern of his kasaya is analysed. Ananda's kasaya is called Saṅghāṭi, inside which is a right lapel wide sleeve shirt（右衽宽袖偏衫）, covered with field pattern robe. It should wrap the left shoulder, bare the right shoulder, folded and naturally drop to the left arm. However, because of the knot connecting the button behind the shoulder, it formed a triangular shape. The painter drew the kasaya shape according to the Buddhist code and the actual production requirements. The kasaya is earth red field pattern with seven stripes on the surface, four petals pattern on the trimming and the leaves (the horizontal bar and vertical bar in the middle). Besides decoration function, the positions of these patterns indicate that they could sew and reinforce the Kasaya.

（Text by: Cui Yan）

图：崔岩 Picture by: Cui Yan

敦煌莫高窟初唐

第71窟

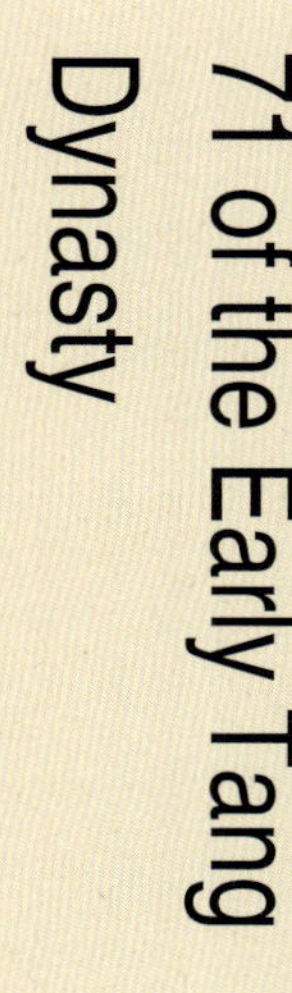

Dunhuang Mogao Grottoes Cave 71 of the Early Tang Dynasty

第71窟为初唐时期创建的覆斗顶洞窟。因此窟曾遭烟熏黑，所以两侧壁面经变画均模糊难辨。后经敦煌研究院以科技保护手段进行部分清洗，得以露出精美的初唐画迹。

窟内顶部四披绘千佛，西壁开一平顶敞口龛，龛内彩塑一佛、二弟子、二菩萨，龛壁浮塑背光，两侧各绘二飞天、赴会佛一铺及项光、莲花，龛顶绘千佛。龛外南、北侧各存孔雀卷草边饰一段。北壁绘阿弥陀经变，南壁绘弥勒经变。

窟内的彩塑和壁画已体现出典型的唐代特征。南北两壁所绘为唐代盛行的经变画，体现了当时流行的净土信仰和弥勒信仰。规模宏大的经变画构图讲究，有序汇聚了数量繁多的人物、建筑、装饰图案等内容，主次分明，虚实相间，体现出画师对于大型经变画技艺的熟练掌握。彩塑和壁画的人物造型也由早期的清秀消瘦渐渐转变为丰满结实，色彩配置大胆且采用退晕法表现立体层次，线条挺拔有力，体现出唐代崇尚强健、富丽的审美倾向。

西壁佛龛内彩塑服饰结构清晰，完好地保留了初唐时期塑像的造型原貌，塑像上图案绘制精美，服装边饰以半团花为主，造型变化丰富，色彩饱和。彩塑服饰目前有很多清晰的细节呈现出来，如菩萨披帛、裙腰的扭转结构、阿难的交领搭叠、袖口和腿部衣裙的褶皱等。北壁上的服饰则较多地保留了印度的服饰形制特征，配色偏爱高饱和度，喜爱使用对比色。

（文：崔岩）

CAVE 71

Cave 71 is a truncated pyramidal ceiling cave built in the early Tang Dynasty. The cave was once blackened by smoke, so the paintings on both sides of the cave were blurred. Later, the Dunhuang Research Academy used scientific method partially cleaned up and revealed the original exquisite paintings of the early Tang Dynasty.

The Thousand Buddha motif is painted on the four slopes of the cave. A flat top open niche carved into the west wall. Inside the niche, it has painted sculptures of one Buddha, two disciples and two Bodhisattvas. The walls of the niche have clay relief of mandorla, two painted flying Apsarases, an attending Buddha, halo, and lotus flower on both sides. The top of the niche is painted with Thousand Buddha motif. On the north and south side of the niche both has a section of peacock scrolling grass pattern. The north wall is painted with Amitabha Sutra illustration and the south wall is painted with Maitreya Sutra illustration.

The colored sculptures and murals in the caves have already reflected the typical characteristics of the Tang Dynasty. The paintings on the north and south walls are the classic paintings popular in the Tang Dynasty, reflecting the popular belief in pure land and Maitreya at that time. The composition of the large-scale sutra tableau painting is exquisite, which orderly gathers a large number of figures, buildings, decorative patterns and other contents. The primary and secondary are clear, and the virtual and real are alternated, which reflects the painter's mastery of large-scale sutra tableau painting techniques. The figures of colored sculptures and murals have gradually changed from delicate and thin to plump and strong. The color configuration is bold, and the three-dimension effect is expressed by the color shading method（退晕法）. The lines are straight and powerful, which reflect the aesthetic tendency of the Tang Dynasty to advocate strength and richness.

The structure of the clothes is clear, and the original appearance of the sculpture in the early Tang Dynasty preserved intact. The patterns on the sculpture are exquisite, and the dress edge is mainly decorated with half flower medallion, with rich modeling changes and saturated colors. At present, there are many clear details on the clothes, such as the Bodhisattva's uttariya, the waist wrap, Ananda's overlapping collar, the folds on cuffs and legs and so on. The clothes on the north wall retains more shape and structure characteristics of Indian clothing, with color matching preferring high saturation and contrasting.

(Text by: Cui Yan)

Contemplating Bodhisattva's clothes

on north wall of the main chamber in cave 71 dated to the early Tang Dynasty at Mogao Grottoes

思维菩萨服饰

莫高窟初唐第71窟主室北壁

图为北壁阿弥陀经变中阿弥陀周围八大菩萨之一，位于主尊左侧，又称为"思维菩萨"。思维菩萨是菩萨造型的一种典型形式，常以右手支颐或扶额作思维状、交脚趺坐或结跏趺坐为标识，在早期犍陀罗佛教造像中已有圆雕或浮雕的先例。敦煌莫高窟早期思维菩萨曾以彩塑形式出现在北凉第275窟、北魏第257、259等窟，常位于象征兜率天宫之圆券龛和阙形龛内。隋代第417窟的思维菩萨悬塑于龛外两侧。绘画形式的思维菩萨至隋、初唐始见于壁画中。

此身菩萨双眸微垂，略泛笑意，面容恬静，侧面低首；右肘承于膝上，支颐沉思；左手轻抚腰间，交脚趺坐于覆瓣莲花座上。画面艺术处理上，人体比例准确，用线笔力遒劲，色彩温润典雅。服饰造型上，菩萨颈间饰璎珞，戴臂钏和手镯，上身斜披披帛，下穿丝质长裙，透明的天衣环绕身体。菩萨梳高髻，长发及肩，头戴宝冠，人物整体造型端稳静谧，突出了菩萨在听法思维时专注澄明的精神境界。身体各部位的装饰物精美灵动，与菩萨平心静思之神态形成动静交互的审美内涵。

（文：刘元风）

The picture shows one of the eight Bodhisattvas around Amitabha on the north wall of the Amitabha Sutra tableau. He is located on the left side of the Buddha, also known as the contemplating Bodhisattva. Contemplating Bodhisattva is a typical manifestation of Bodhisattva; He often uses the right hand to support his cheek or to hold the forehead as the thinking posture, and sit with ankles crossed or legs folded. There are precedents of round sculpture or relief sculpture in the early Gandhara period Buddhist sculptures. The early contemplating Bodhisattvas of Mogao Grottoes used to appear in the form of colored sculptures in caves 275 of the Northern Liang Dynasty, 257 and 259 of the Northern Wei Dynasty, and were often located in the round edge niches and Que (watch tower) shaped niches, which symbolize the Tusita Heaven. The contemplating Bodhisattva in cave 417 of the Sui Dynasty are suspended on both sides of the niche. The painting form contemplating Bodhisattva was first seen in the Sui and early Tang Dynasties.

The Bodhisattva's eyes are slightly drooping, slightly smiling. His face is quiet, head is bowed, and the right elbow on his knee. The left hand rests on the waist, the feet crossed and sit on the lotus seat. In terms of artistic style, the body's proportion is accurate, lines are strong, and colors are warm, soft and elegant. In terms of dress modeling, the Bodhisattva is decorated with keyūra, armlets and bracelets. The upper body is draped with uttariya, and the lower part is wearing a silk antariya, which seems to be made in heaven and almost transparent. The Bodhisattva has a high bun, long hairs covering the shoulders, and a crown on his head. The overall shape of the Bodhisattva is stable and quiet, which highlights the spiritual state of the Bodhisattva when listening to the Dharma. The ornaments of various parts of the body are exquisite and flexible; the expression of Bodhisattva's meditation is peaceful, which forms a aesthetic connotation of dynamic and static interaction.

(Text by: Liu Yuanfeng)

图：刘元风　Picture by: Liu Yuanfeng

供养菩萨服饰

莫高窟初唐第71窟主室北壁

The offering Bodhisattva's clothes

on north wall of the main chamber in cave 71 dated to the early Tang Dynasty at Mogao Grottoes

图为北壁阿弥陀经变中阿弥陀周围八大菩萨之一，位于主尊左下方，又称“胡跪菩萨”。慧琳所作《一切经音义》中说：“胡跪，右膝着地，竖左膝危坐，或云互跪。”这种姿势又被称为“左跪”，即右膝和右足着地，竖左膝，左足踏地，佛经中提到的“右膝着地者为正仪”即是指胡跪，如倦累可两膝姿势互换，是一种表示敬意的姿势。这本是西域少数民族半蹲半跪的姿态，后来演变为佛教礼节。敦煌莫高窟西魏第285窟南壁的五百强盗成佛因缘故事画中，五百强盗在成佛时便是以此姿势跪于佛陀座前表示皈依和觉悟。

此身菩萨神情安静优雅，梳高髻，长发分披两肩，头戴宝冠，挺胸昂首，左手轻拈颈间璎珞，右手抬举于胸前，拈一朵蒂部翻卷、冠瓣饱满的宝相花苞供养，侧身胡跪于覆瓣莲花座上。菩萨上身袒裸，下身穿红色长裙，腰系粉绿色围腰，身上环绕天衣。在画面的艺术处理上，线条简练流畅，表现其薄衣轻纱的质感和衣装丝织品的精良。人物服饰色彩对比鲜明，配置协调，整体造型稳重端庄，体现了菩萨恭敬供养、静中欲动的神态。

（文：刘元风）

The picture shows one of the eight Bodhisattvas around Amitabha on the north wall of Amitabha Sutra tableau. He is located at the lower left of the main image, also known as Hu kneel (one knee bent) Bodhisattva. Huilin wrote in *Yiqiejingyinyi* (《一切经音义》), a dictionary of difficult vocabulary in sutra) said: “Hu kneel, right knee on the ground, erect left knee sit firmly, also called alternate kneel.” This posture is also known as left kneel, that is, the right knee and the right foot on the ground, erect the left knee, and step on the ground with the left foot. The Buddhist Scripture recorded that “the one who pud down the right knee is the proper way” refers to Hu kneel. If someone feels tired, the posture of the two knees can be exchanged; this posture is a gesture of respect. Originally this is the posture of half squatting and half kneeling that was popular among ethnic minorities in the Western Regions, and later it is adapted into Buddhist manner. In the story painting of the five hundred bandits who became Buddhas on the south wall in cave 285 of the Western Wei Dynasty in Mogao Grottoes, the five hundred robbers used Hu kneel in front of Buddha seat to express their conversion.

This Bodhisattva looks quiet and elegant. He fastened a high bun and long hairs on his shoulders. He wears a treasure crown on his head and holds head high. His left hand gently pinches the necklace. His right hand is raised in front of his chest holding up a precious flower bud with full petals. He kneels on the lotus seat by one kneel. The Bodhisattva's upper body is naked, the lower body is wearing a red antariya, the waist wrapped by a piece of pink green cloth, and the body is surrounded by heavenly clothes. In the artistic treatment of the picture, the lines are concise and smooth, showing the texture of the light and thin cloth and the fine silk fabrics. The clothes of the character has distinct color contrast, coordinated configuration, and the overall shape is sedate and dignified, which reflects the attitude of the Bodhisattva's reverence and devotion and the tendency to move out of stillness.

(Text by: Liu Yuanfeng)

图：刘元风 Picture by: Liu Yuanfeng

The pattern on painted sculpture Bodhisattva's waist wrap

on the north side of the west niche in the main chamber of cave 71 dated to the early Tang Dynasty at Mogao Grottoes

彩塑菩萨裙腰图案

莫高窟初唐第71窟主室西壁龛内北侧

第71窟西壁龛内北侧彩塑菩萨的裙腰图案仅露出中央一个完整单体花型，以红底色衬托青、绿色花纹，主体花型由花叶、花梗、花托、花萼构成。花托与萼片相融合处理成波浪状左右散开，动势强烈，其上正中绘制三瓣花，后层复点缀两片较小花瓣，增加层次感，以此为中央向左右两侧不同方向添加流畅的变形花萼，花萼之上点缀大小不一的花瓣若干。纹样组织配置十分灵活，可见唐代画师构思巧妙，将花朵从自然形中脱胎出来，加以变形打散重构，使其平面化、装饰化，赋予了装饰花型以初唐的时代特征与气魄。

（文：高雪）

The pattern on the waist wrap of the painted sculpture Bodhisattva on the north side of the west niche in cave 71 only shows a complete single flower pattern in the center, with the blue and green pattern complement by the red background. The main flower design is composed by leaves, pedicels, receptacle and calyx. Three petals were drawn in the center of the flower, and two smaller petals were interspersed on the outer layer to increase the sense of depth. Smooth transformed calyxes were added to the left and right sides in different directions from the center, and several petals of different sizes were dotted on the calyx. The arrangement of pattern is very flexible; it can be seen that the painters of the Tang Dynasty had ingenious ideas. They took the flowers out of the natural form and transformed them, scattered them and reconstructed them, making them flat and decorative, giving the decorative patterns the characteristics and vigor of the early Tang Dynasty.

(Text by: Gao Xue)

The pattern on painted sculpture Buddha kasaya's trimming and antariya trimming

in the west niche of the main chamber in cave 71 dated to the early Tang Dynasty at Mogao Grottoes

彩塑佛陀袈裟缘边、裙边图案

莫高窟初唐第71窟主室西壁龛内

第71窟的佛陀袈裟下摆图案用半破式二方连续的宝相花衬托佛陀的庄严。宝相花是由自然形态的花卉抽象概括而得的一种植物造型，多呈对称放射状。它把盛开的花、蕾、叶等组合，形成更为平面化、更具装饰性的图案。这里的宝相花以红、绿、蓝、赭石、白为主色，庄重艳丽，两半交错的排列在空间上有一定的起伏节奏感，具有典型的初唐特征。花瓣仍保持着莲花的特征，边缘简洁，这与该洞窟中弟子阿难的头光以及佛龛外沿的边饰花型相一致。在织锦技术上，唐代一改汉魏六朝经线起花的技术特点，用两组或两组以上的纬线同一组经线交织，以纬锦工艺使织物色彩更加丰富多样，同时也为细腻繁复的宝相花纹样在唐代的流行提供了纺织技术支持。

裙缘二方连续图案以半破式的如意纹为中心，以辐射状表现双层花瓣，花瓣的平瓣外形曲线节奏平缓，内部的如意纹流畅优美且空间较大，花瓣左右相连、内外重叠，整体呈现出向外开放的张力。

（文：高雪）

The two consecutive half Baoxiang flower pattern on the bottom trimming of Buddha's Kasaya in cave 71 complements the Buddha's solemnity. Baoxiang flower is a kind of plant shape which was reconstructed abstractly from natural flowers. They are mostly symmetrical and radial. They combined the blooming flowers, buds and leaves to form a more plane and decorative pattern. The Baoxiang flowers here are mainly red, green, blue, ochre and white. They are solemn and gorgeous, arranged alternately in two parts, which have a certain sense of ups and downs in space, showing the typical characteristics of the early Tang Dynasty. The petals still retain the characteristics of lotus flower, with clean edges, consistent with the head light of disciple Ananda in the cave and the decorative flower patterns on the outer trimming of the niche. In terms of brocade technology, in the Tang Dynasty people developed the waving technology from the Han, Wei and Six Dynasties, they started using two or more groups of weft threads to interweave one group of warp threads. The weft brocade technology made the fabric more colorful and diversified. At the same time, it also provided technical support for the popularity of delicate and complicated Baoxiang flower pattern in the Tang Dynasty.

The two in a group repeated pattern of Kasaya's trimming is centered on the half broken Ruyi (an S-shaped ornamental object, a symbol of good luck) pattern, which shows double-layer petals in a radial manner. The flat petal's curves are gentle, and the inner Ruyi pattern with large space is smooth and beautiful. The petals are connected from left to right and overlapped inside and outside, showing the outward tension.

(Text by: Gao Xue)

图：高雪　Picture by: Gao Xue

图：高雪　Picture by: Gao Xue

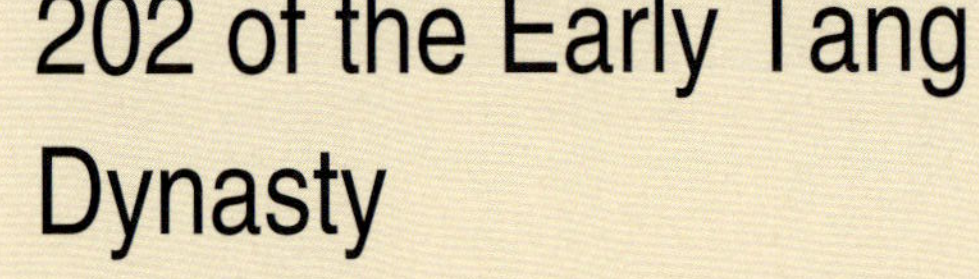

敦煌莫高窟初唐
第202窟

Dunhuang Mogao Grottoes Cave 202 of the Early Tang Dynasty

第202窟是始建于初唐时期的方形覆斗顶窟，创建时完成了西壁龛内的绘塑和东壁门上的说法图，其余诸壁完成于中唐，至宋代时重修了窟门，并补绘窟顶和甬道的壁画。主室西壁开斜顶敞口龛，龛内有初唐彩塑趺坐佛像一身，弟子、菩萨、金刚力士各二身，均经清代重修。龛顶绘法华经的见宝塔品，龛内西壁中唐改画背光，南北两侧初唐绘弟子八身、菩萨二身。从整体的色彩搭配和人物造型中能够感受到初唐时期清丽的艺术风格。

本窟西壁龛顶和龛内壁及东壁门上存有几组初唐人物，从服装款式上来看，菩萨多为上身斜披帛，下身着半透明长裙搭配璎珞和飘带，色彩疑为后期补绘，弟子多着土红色袈裟。

（文：崔岩）

Cave 202 is a truncated pyramidal ceiling cave built in the early Tang Dynasty. When it was carved, the painted sculptures in the west niche and the painting on the east wall above the door were completed. The rest of the walls were completed in the middle Tang Dynasty. In the Song Dynasty, the cave gate was rebuilt and the murals on the top of the cave and the corridor were painted. The west wall of the main chamber has an open niche with a sloping top. Inside the niche, there is a painted sculpture of a sitting Buddha with legs folded made in the early Tang Dynasty and two bodies of disciples, Bodhisattvas and Vajra warriors, all of which were repaired in the Qing Dynasty. The top of the niche is painted with The Emergence of Treasured Pagoda from Lotus Sutra, and the west wall inside the niche is painted with backlight in the middle Tang Dynasty, beside them are eight bodies of disciples and two bodies of Bodhisattvas painted in the early Tang Dynasty. From the overall color matching and character modeling, we can feel the clean and beautiful artistic style of the early Tang Dynasty.

There are several groups of figures of the early Tang Dynasty on the top of the west niche, the niche walls and above the east door. From the perspective of clothes style, the Bodhisattvas mostly wear uttariya on the upper body, and wear translucent long antariya with wreaths and ribbons on the lower body. The colors were probably added up in the later period, and the disciples mostly wear earthen red kasayas.

(Text by: Cui Yan)

弟子像袈裟图案

莫高窟初唐第202窟西壁龛内北侧

The pattern on disciple's kasaya

on the north side of the west niche in cave 202 dated to the early Tang Dynasty at Mogao Grottoes

这里选取的是此窟西壁龛内北侧一身弟子像的袈裟图案。从壁画中可以看到，袈裟按照佛典规制进行绘制，装饰以石青、石绿和褐色的横向斑纹图案，并以虚线示意竖向的绗缝针法。从肌理表现手法来看，此图案似为文献中记载的树皮袈裟。根据敦煌文献S.6208《杂集时要用字》所载“布部：火麻。高机。树皮。单纻。土纻。”可知“树皮”是一种面料的称谓，早在隋代时已用于对袈裟的称呼。如隋炀帝《入朝遣使参书》中记载：“弟子总持和南：垂赐万春树皮袈裟一缘，……谨寻菩萨戒称所著袈裟，皆染使坏色。”日本正仓院中藏有九领“御袈裟”（八世纪），其中有两条名为“九条刺纳树皮色袈裟”和“七条刺纳树皮色袈裟”，以各种杂色绫锦缀成纹样，色彩犹如树皮，与此图案十分相似。可见，敦煌壁画中的服饰图案在历史真实来源方面具有一定的可靠性，是值得深入研究的珍贵图像资料。

（文：崔岩）

Here is the disciple's kasaya pattern on the north side of the west niche in this cave. From the mural, we can see that the kasaya was painted according to Buddhist principle, decorated with azurite, malachite green and brown horizontal pattern, and the dotted lines indicate the vertical quilting needle technique（绗缝针法）. From the aspect of texture expression, this pattern seems to be the bark kasaya recorded in the literature. According to Dunhuang document S.6208 *Zajishiyaoyongzi*（杂集时要用字）, the cloth chapter: “Cannabis（火麻）, Gaoji（高机）, Bark（树皮）, Single Twine（单纻）, Earth Twine（土纻）.” It can be seen that “bark” was a kind of fabric appellation, and as early as the Sui Dynasty it has been used to address the kasaya. For example, in the Sui Yang emperor's *Envoy records in court*（入朝遣使参书）of the Sui Dynasty, it was recorded that “disciple Zongchihenan（总持和南）bestow the bark kasaya as a gift... according to Bodhisattva's precepts, kasaya is dyed with bad colors.” There are nine “Royal kasaya”（御袈裟，8th century）collected in the Japanese Shoso-in. Two of them are named “nine cina bark color kasaya（九条刺纳树皮色袈裟）” and “seven cina bark color Kasaya（七条刺纳树皮色袈裟）”. They are decorated with various kinds of variegated damask brocade. The color is like bark, which is very similar to this pattern. It can be seen that the clothes patterns in Dunhuang murals have certain reliability, and are valuable image materials worth in-depth study.

（Text by: Cui Yan）

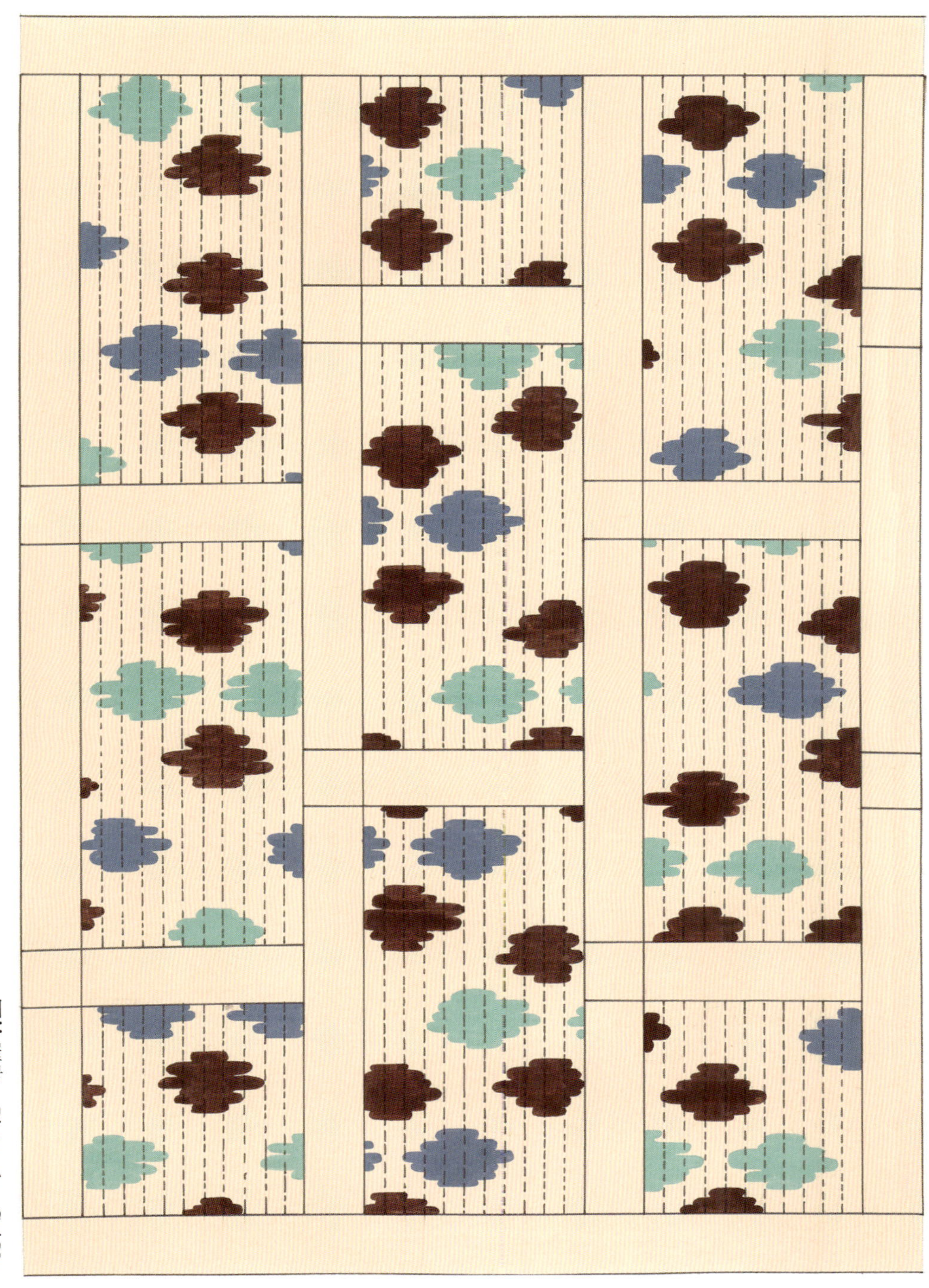

图：崔岩 Picture by: Cui Yan

敦煌莫高窟初唐

第204窟

Dunhuang Mogao Grottoes Cave

204 of the Early Tang Dynasty

第204窟为初唐时期开凿的覆斗顶洞窟，也有学者认为开凿时间可能早至隋末。主室窟顶藻井为飞天莲花井心，四披画千佛。主室西壁开重层龛，龛内彩塑一佛、二弟子、四菩萨，有的塑像经宋代和清代重修。内层龛壁中央浮塑火焰纹背光，顶部绘飞天和骑龙天人，两侧绘二弟子、八菩萨、婆薮仙、鹿头梵志等。外层龛绘火焰莲花化生龛楣，两侧为飞天和供养菩萨。龛外绘千佛，龛下绘十身供养菩萨和二身供养童子。主室南北壁千佛中央绘说法图。东壁门上和南北两侧绘千佛，并存有宋绘花卉，以及供养菩萨五身。前室和甬道残存五代和宋代壁画。

此窟中的壁画和彩塑明显带有隋代遗风，如内层龛壁在土红地上绘制的飞天和骑龙天人图中流云飞动，与隋代龛顶的装饰有相近之处；龛内所塑两尊供养菩萨较为直立的身躯和联珠纹裙饰，也显示出隋代塑像和装饰艺术的特点。但仔细观察可以发现此窟的飞天形象更加世俗化，外层龛北侧彩塑菩萨虽然体型仍显粗壮，但她略微倾斜的身躯、丰润的脸庞和颔首的神情都透露出丰肌秀骨的特色，说明在隋代原有的艺术传统中已经萌发出新的唐代风尚。

相较于南、北、西三壁彩塑和壁画中人物服饰的工整之风，四壁下部供养菩萨身姿灵动，有着中原交领长袍者，但大多为僧祇支和披帛搭配长裙，披帛飞扬，整体一派轻盈飘逸之气。

（文：崔岩）

DUNHUANG MOGAO GROTTOES

CAVE 204

Cave 204 was excavated in the early Tang Dynasty with a truncated pyramidal ceiling. Some scholars think that the excavation time may be as early as the end of Sui Dynasty. In the center of the caisson ceiling of the main chamber there is a louts flower, with Thousand Buddha motif painted on the four slopes. In the west wall of the main chamber are double layers niche, in which has one Buddha, two disciples and four Bodhisattvas painted sculptures. Some of them were repaired in the Song Dynasty and Qing Dynasty. The center of the inner niche wall is decorated with flame pattern backlight relief. The top is painted with flying Apsaras and man riding on dragon. On both sides are painted two disciples, eight Bodhisattvas, Posou immortal and Lutou Brahman. The outer niche is painted with flames, lotus and incarnations on the edge, flying Apsarases and Bodhisattvas on both sides. The outside of the niche is painted with Thousand Buddha motif, and under the niche there are ten bodies of offering Bodhisattvas and two bodies of offering children. In the center of the Thousand Buddha motif on the north and south walls of the main chamber, a preaching scene is drawn respectively. There are Thousand Buddha motif painted above, north and south sides of the east gate. There are also flowers and five bodies of offering Bodhisattvas painted in Song Dynasty. The vestibule and corridor have murals of Five Dynasties and Song Dynasty.

The murals and colored sculptures in this cave clearly bear the Sui Dynasty legacy. For example, the clouds flying in the sky and man riding on the dragon with red earth background in the inner niche walls are similar to the decoration on the niche top in the Sui Dynasty. The two upright bodies and beaded antariya ornaments in the niche also show the characteristics of the sculptures and decorative arts in the Sui Dynasty. However, a close observation shows that the image of flying Apsaras in this cave are more secular. Although the figure of the painted sculpture Bodhisattva on the north side of the outer niche is still robust, his slightly inclined body, round face and chin, the expression all reveal the plump body and thin bones, which indicate that a new Tang Dynasty fashion had sprouted in the original artistic tradition of the Sui Dynasty.

Compared with the neat clothes style of the painted sculptures and murals of the three walls in the south, north and west, Bodhisattvas at the lower part of the four walls have flexible postures, and some of their robes are from the Central Plains, but most of them wear Sankaksika, uttariya and antariya. The uttariya flies in the air, giving an overall atmosphere of light and elegant.

(Text by: Cui Yan)

彩塑菩萨裙身图案 莫高窟初唐第204窟主室西壁龛

The pattern on painted sculpture Bodhisattva's antariya on the west niche of the main chamber in cave 204 dated to the early Tang Dynasty at Mogao Grottoes

第204窟主室西壁内层龛南侧的彩塑菩萨着袒右僧祇支，束围腰，下着长裙。裙子以浅土红为底色，上绘竖条装饰带，主体图案为联珠纹。联珠纹外圈以浅石青为底色，中央为白线勾勒的十二瓣小团花纹，花心的圆圈与周围的联珠相呼应。每个图案单元竖向延伸排列，其间将菱形散射状花蕊破开填充在两个团花联珠纹中间，规律搭配并穿插有致。

联珠纹起源于萨珊波斯，常以狩猎、翼马、野猪头、大角鹿为题材，后经西域传入中国，是北朝至唐代重要的流行图案之一。在新疆吐鲁番阿斯塔那墓曾发现联珠孔雀纹锦、联珠对马纹锦等纺织品实物，在敦煌石窟隋代的壁画和彩塑中联珠纹也屡见不鲜，并发展出凤纹、花卉等喜闻乐见的主题，说明外来文化沿丝绸之路的广泛传播和本土化融合。

（文：王可、崔岩）

The painted sculpture Bodhisattva on the south side of the inner niche in the west wall of cave 204 wears Sankaksika and bares the right shoulder, girdle around his waist and a long antariya on the lower body. The antariya used light earth red as the background color, and painted with vertical stripes, and the main body pattern is Sassanian roundels. The outer ring of the roundels pattern has light azurite background, and the center is a twelve petals small flower medallion pattern outlined by white lines. The circle of the flower center echoes with the surrounding beads. Each pattern unit is arranged vertically; the scattered rhombus shaped stamens are broken and filled in the middle of the two flower medallions and beads pattern, which are arranged regularly and interspersed.

Sassanian roundels pattern originated from Sassanian Persian, often appeared with hunting scene, winged horse, wild boar head, bighorn deer and so on. Later it was introduced into China through the Western Regions and became into one of the important popular patterns during the Northern Dynasty to the the Tang Dynasty. In the Astana tomb in Turpan, Xinjiang, some textile objects, such as Sassanian roundels peacock pattern brocade（联珠孔雀纹锦）, Sassanian roundels double horse pattern brocade（联珠对马纹锦）are preserved, at the meantime, Sassanian roundels pattern also have been found in murals and colored sculptures in Dunhuang Grottoes dated to the Sui Dynasty, it even developed phoenix pattern and flower pattern, which indicate the wide spread of foreign culture along the Silk Road and the integration of different cultures both at home and abroad at that time.

（Text by: Wang Ke, Cui Yan）

彩塑迦叶袈裟图案 莫高窟初唐第204窟主室西壁龛内

The pattern on painted sculpture Kasyapa's Kasaya in the west niche of the main chamber in cave 204 dated to the early Tang Dynasty at Mogao Grottoes

第204窟主室西壁龛内彩塑弟子像虽然经过后世重修，但是服饰穿搭和图案还保留了部分初唐风貌。二弟子均外披袒右田相纹袈裟，内为土红色袈裟，下着裙。最外层袈裟的图案十分特别，整体为石青色田相纹。由于律典规定制作三衣的布料皆为“坏色”（又名“浊色”，即袈裟色），所以坛（田相格）内用晕染和涂抹黑色不规则斑点的方式模拟“坏色”的效果，并用虚线模拟缝纫的线迹。缘边（四周边缘）与叶（中间的横条和竖条）上没有装饰，用纯色布条缝缀，体现了袈裟本有的朴素质地。

（文：王可）

Although the painted sculptures of disciples in the west niche of the main chamber in cave 204 have been repaired by later generations, their clothes and patterns still retain some features of the early Tang Dynasty. The two disciples all wear kasaya bared the right shoulder with the field pattern on the outside, earth red kasaya inside. The pattern of the outermost kasaya is very special. The overall pattern is azurite field pattern. As the Buddhist precepts require that the cloth used to make kasaya has to be bad color (also known as turbid color, that is, kasaya color) , the effect of “bad color” is simulated by shading and smearing irregular black spots in the altar (the squares on kasaya), and the sewing thread is simulated by dotted lines. There is no decoration on the trimming (clothes'edges) and leaves (the horizontal and vertical bars in the middle). They are sewn with solid color cloth, reflecting the simple texture of kasaya.

（Text by: Wang Ke）

图：王可　Picture by: Wang Ke

图：王可　Picture by: Wang Ke

敦煌莫高窟初唐

第220窟

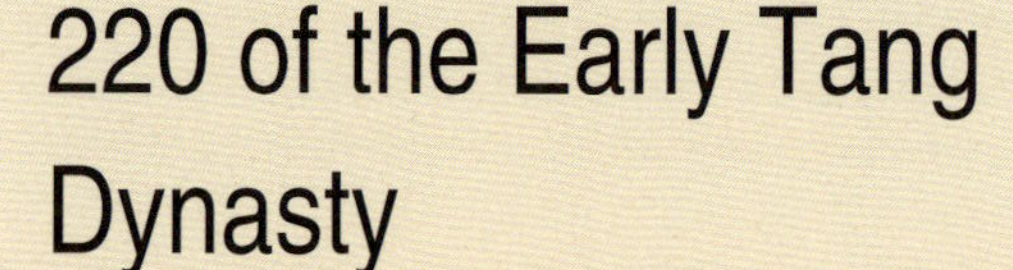

Dunhuang Mogao Grottoes Cave 220 of the Early Tang Dynasty

第220窟是创建于初唐时期的覆斗顶洞窟，后经中唐、晚唐、五代、宋、清重修，是唐代的代表窟之一。宋或西夏时，窟内壁画全被覆盖，绘以千佛。1943年，四壁之上层壁画被剥开，使初唐壁画杰作赫然重晖。因主室西壁龛沿下有“翟家窟”题记遗存，以及东壁门上有贞观十六年（642年）题记一方，所以此窟又被称为“翟家窟”或者“贞观十六年窟”。

此窟主室西壁开龛，内彩塑一佛、二弟子、二菩萨，经清代重修，龛外两侧绘文殊变、普贤变各一铺。主室南壁绘无量寿经变，以天空、水面、地面三大空间构图表现西方极乐世界不鼓自鸣、亭台水榭、歌舞升平等种种盛景，是莫高窟最早、场面最大的净土变。主室北壁绘药师经变，主体为七身药师佛立像、胁侍菩萨等。经变下部有巨大的灯轮，两侧有四位舞者相对起舞，是敦煌壁画中著名的乐舞图之一。东壁门上绘说法图一铺，男女供养人各一身并题记一方；门两侧画维摩诘经变。维摩诘经变场面宏大，形象塑造生动逼真，其中帝王群臣像可与唐代阎立本所绘《历代帝王图》媲美，其人物造型和艺术手法如出一辙，反映了敦煌壁画在中国古代绘画史上的重要地位和历代帝王图式的流传影响。

西龛内塑像的服饰大体保持了初唐的原貌，但是色彩已经被重绘。西龛内西壁弟子袈裟的纹样有山水、团花等多种题材，西龛内南北两壁的菩萨和东壁门上、南北两壁等多尊菩萨服饰绘制清晰、色彩饱满，其中尤以南壁最为丰富。南壁无量寿经变中菩萨服饰以上衣斜披帛和裙的搭配为主，喜爱表现青绿色调的半透明的面料效果，尤其是或蓝或绿的披帛穿插搭叠在人物和莲花上，其上还绘有小型图案，疑为仿唐时印花或绞缬效果。小菱格纹是当时非常受欢迎的纹样，在多处裙子上都有所表现，只是颜色和具体的造型有所差别。东壁维摩诘经变中各国王子和国王大臣形象展示了当时唐代社会世俗服饰的样貌。帝王服饰的冕旒、肩挑日月，大臣的交领长袍，以及王子的帽饰、圆领长袍，武士的铠甲，仪仗人员的裲裆等，形象都非常清晰完整。北壁佛衣中出现了类似石榴裙的底摆样式，下部跳胡旋舞和演奏的乐队呈现了褶皱细密、边缘翻卷的石榴裙和飞扬的披帛绸带，方格或斜向菱格纹搭配联珠、半团花边饰为初唐时期主要的服饰装饰纹样。

（文：崔岩）

Cave 220 is a cave built in the early Tang Dynasty with a truncated pyramidal ceiling; it was repainted in the middle Tang Dynasty, late Tang Dynasty, Five Dynasties, Song Dynasty and Qing Dynasty, which is one of the represent caves of the Tang dynasty. In the Song or Xixia period, the murals in the cave were repainted and covered by Thousand Buddha motif. In 1943, the upper layer of the four walls was stripped off, which made the masterpiece of the early Tang Dynasty's murals shine again. The cave is also known as "Zhai Family Grotto" or "the sixteenth year of Zhenguan Grotto" because of the remains of family name below the west niche of the main chamber and the inscription of the 16th year of Zhenguan (642) above the east wall gate.

In the west wall of the main chamber of the cave is a niche with painted sculptures of one Buddha, two disciples and two Bodhisattvas. After been repaired in the Qing Dynasty, the both outsides of the niche are painted with Manjusri tableau and Samantabhadra tableau. The south wall of the main chamber is painted with Aparimitayur-sutra tableau. The composition of the sky, water and ground shows the magnificent scenes of the Western Paradise, such as self-playing instruments, waterside pavilions, singers and dancers, etc. It is the earliest and largest pure land illustration in Mogao Grottoes. The north wall of the main chamber is painted with the illustration of Bhaisajyaguru sutra. The main images are the portrait of the seven bodies of Bhaisajyaguru Buddha and the attendant Bodhisattva. There is a huge lamp wheel at the bottom of the tableau, four dancers on both sides. It is one of the most famous music and dance scenes in Dunhuang murals. Above the gate on the east wall is painted a preaching scene, one male donor and one female donor who both have inscription respectively. On both sides of the door are painted Vimalakirti Sutra illustration, Vimalakirti's image is vivid and lifelike. The images of emperor and ministers are comparable to the painting *The Emperors of Past Dynasties* by Yan Liben of the Tang Dynasty. Their figure style and artistic technique are the same, reflecting the important status of Dunhuang murals in the history of ancient Chinese painting and the influence of the spread of the emperor painting.

The clothes of the sculptures in the west niche have largely kept the original appearance of the early Tang Dynasty, but the colors have been repainted. The patterns on disciples' kasaya in the west niche include landscapes, flower medallion and other themes. Clothes of Bodhisattvas in the north and south side of the west niche, above the east door and on the north and south walls are clearly painted and full of colors, among which the south wall is the most abundant. The Bodhisattva dress in Aparimitayur-sutra tableau on the south wall is mainly the combination of the uttariya and antariya, which likes to show the green color translucent fabric effect, especially the blue and green uttariya interspersed and overlapped on the figures and lotus flowers, and there are also small patterns painted on them, which are supposed to imitate the effect of printing or weaving in the Tang Dynasty. The rhomboid pattern was a very popular pattern at that time, and it appeared on many dresses, but the color and the specific shape varied. The images of princes, kings and ministers in the Vimalakirti sutra tableau on the east wall show the secular clothes of the Tang Dynasty. The image of the emperor's dress is very clear and complete, such as the Diadem with Tassels (冕旒) of the emperor, the sun and the moon on his shoulders, the minister's cross collar robe, the prince's hat and round collar robe, the warrior's armor, and the Liangdang (裲裆) of the guard of honor. The Buddha's clothes on the north wall has the bottom of the pomegranate skirt style. At the lower area, the Hu-swirling dance and the performing band have the pomegranate skirt with fine folds and rolling edges and flying silk ribbon. The squares or diagonal rhomboid pattern with beads and half flower medallion are the main clothing decoration patterns in the early Tang Dynasty.

(Text by: Cui Yan)

伎乐人服饰

莫高窟初唐第220窟主室南壁

The Dancer's clothes

on the south wall of the main chamber in cave 220 dated to the early Tang Dynasty at Mogao Grottoes

此为初唐第220窟主室南壁无量寿经变下部乐舞场景中的舞蹈天人之一。舞者头部微侧，目光视下。左手上扬，擎举长巾，右手紧握长巾，做下压之势。跣足，左腿立于装饰有联珠纹的圆毯之上，右腿屈膝抬起，表现了舞蹈瞬间富有动感的姿态。另一舞者与其相对，姿态相向而类似，在乐队的伴奏下翩翩起舞。

舞者头戴宝冠，梳高髻，辫发飘散，戴璎珞、臂钏和手镯。上身斜披披帛，胯间有围腰装饰，下着十字花纹荷叶边阔腿裤，其类似裙裤的设计不仅造型时尚而且利于舞者的肢体动作，是一种独具特色的舞蹈服装。舞者手中长巾上下翻飞，身上的璎珞、飘带和腰带随风而动，增加了人物的舞蹈动势，集中呈现了这种源自西域的舞蹈艺术的风格与服饰特色。整体服饰色彩以对比色为主，头冠、披帛和围腰为石青色或石绿色，裤子为具有透明质感的土红色，而在裤脚荷叶边翻起的反面又点缀零星石绿色，长巾和腰带的正反两面也以红绿二色对比配置，以富于张力的艺术手法突出表现了舞者在整幅经变画中的醒目地位。

（文：刘元凤）

This is one of the heavenly dancers located in the lower part of Aparimitayur-sutra tableau on the south wall of cave 220 dating back to the early Tang Dynasty. The dancer's head is slightly sloping, looking down. She is raising the left hand, holding up the long ribbon, and the right hand is grasping the long ribbon tightly to make a downward momentum. The feet are bare; the left foot stands on the round carpet decorated with beads pattern, and the right leg is raised with knee bent, which shows the dynamic posture of dancing movement. The other dancer, opposite to her, is dancing with the accompaniment of the band.

This dancer wears a crown, a bun, braids, and bracelets. The upper body is draped with oblique uttariya. The waist is decorated by waist wrap, and the cross pattern lotus leaf trimming wide leg pants on the lower body. The design which is similar to the pantskirt is not only fashionable, but also convenient to the body movements for dancers, it is a kind of special dance clothes. The dancers' long ribbons fly up and down, and her necklace, ribbons and belt move with the wind, which increase the dancing momentum of the character, and also manifested the style and dress characteristics of this dance art originated from the Western Regions. The color of the whole dress is mainly contrast color. The head crown, uttariya and waist wrap are azurite or malachite green. The trousers are earthy red with transparent texture. The inside of the trousers leg ruffles are decorated with scattered malachite green. The front and back sides of the long scarf and belt are also matched with red and green colors, which highlight the dancer's striking position in the whole painting with rich artistic techniques.

(Text by: Liu Yuanfeng)

图：刘元风 Picture by: Liu Yuanfeng

供养菩萨服饰

莫高窟初唐第220窟主室南壁

The offering Bodhisattvas's clothes

on south wall of the main chamber in cave 220 dated to the early Tang Dynasty at Mogao Grottoes

这两身供养菩萨来自初唐第220窟主室南壁无量寿经变左侧下部，是极乐世界中围绕在阿弥陀佛周围听法的圣众。二身供养菩萨均跣足立于仰瓣莲花座上，前面一身菩萨侧面向前，左手上指，右手捧持柄香炉，左脚足跟着地，足尖翘起，似乎在和着伎乐人的音乐和舞姿打着节拍；后一身菩萨回首顾盼，左手持柄香炉，右手在胸前做拈花状，表现了她们聆听佛法、恭敬供养的情景。

二身供养菩萨的服饰造型较为简洁明晰，前一身菩萨在头后挽大髻，辫发披肩，后一身菩萨梳高髻，均戴宝珠头冠、臂钏、手镯或颈饰。二菩萨上身斜缠披帛，腰间束石绿色围腰，下身着石青色和褐色短裙，并分别穿着菱格四簇点纹和十字花纹印花裤。其裤型均为阔腿缩口，似灯笼形，面料质感轻薄透明。这种围腰套穿短裙和阔腿裤的组合是敦煌唐代壁画所绘菩萨像中常见的服饰之一，初唐第332窟、第329窟均有类似的穿搭表现，裤子通常用透明的面料制作和表现，极具异域风情。在画面艺术处理上，线条飘逸，色彩以黑赭配石青、石绿，格调淡雅。

（文：刘元风）

These two bodies offering Bodhisattvas from the lower left side of the Aparimitayur-sutra tableau on the south wall of cave 220 dated back to the early Tang Dynasty. They are the beings who listen to Dharma around Amitabha in the paradise. The two Bodhisattvas are standing on the lotus seat with bare feet. The front Bodhisattva stands on the lotus seat with his left hand pointing forward and holding the censer in his right hand. His left foot is on the ground with toes tilted up; it seems that the Bodhisattva is tapping the rhythm beats of the dancers and musicians. The second Bodhisattva is looking back, holding the censer in the left hand and performing flower gesture before his chest by the right hand; all these are trying to show they are listening dharma carefully and offering Buddha respectfully.

The dress styles of the two Bodhisattvas are relatively simple and clear. The former one wears a big bun at the back of his head and braided hairs cover the shoulders, while the latter one has a high bun, wearing a jewel crown, armlets, bracelets and neck ornaments. The two Bodhisattvas wear uttariya on the upper body, a malachite green waist wrap around the waist, an azurite skirt and a brown skirt on the lower body, and a rhombus lattice four cluster dot pattern and cross pattern printed trousers respectively. The pants are wide but tight at the openings, like lantern shape; the fabric texture is light and transparent. This kind of combination of short skirt and wide leg trousers is one of the common clothes for Bodhisattvas painted in Dunhuang murals in the Tang Dynasty. Cave 332 and 329 of the early Tang Dynasty have similar clothing style. The pants are usually made and expressed by transparent fabric, which are very exotic. In the artistic treatment of the picture, the lines are elegant, the colors are black ochre with malachite green and azurite, and the style is simple and noble.

(Text by: Liu Yuanfeng)

图：刘元风 Picture by: Liu Yuanfeng

The offering Bodhisattva's clothes

on the north wall of the main chamber in cave 220 dated to the early Tang Dynasty at Mogao Grottoes

供养菩萨服饰

莫高窟初唐第220窟主室北壁

此处选取的是初唐第220窟主室北壁药师经变中的供养菩萨。菩萨位于中央药师佛足下莲花台的栏杆前，跪坐于从池水中生长出的覆瓣莲花宝座之上，似为悬空而置，表现了药师佛所在东方净琉璃世界的美妙殊胜。她仰首注目上方的主尊，面容安静肃然，眉眼细长，长耳垂肩。左手持莲花花苞，右手支撑在身后的莲花座上，姿态优美，神情专注。

在服饰造型上，菩萨头后挽大髻，辫发飘扬，戴宝珠头冠、耳环、颈饰、臂钏、手镯，首饰上均镶嵌着蓝色或绿色的宝珠，并用退晕法表现出晶莹剔透的质感，精致考究。菩萨上身袒裸，环绕天衣，下着长裙，材质轻柔飘拂。人物姿态优美，用线生动传神，色彩沉稳而素雅，图案精美雅致，充分彰显了初唐壁画的典型艺术特征。

（文：刘元风）

Here we choose the offering Bodhisattva who is on the north wall of the main chamber of cave 220 dating back to the early Tang Dynasty. The Bodhisattva is located in front of the railings of the lotus platform under the feet of the central Bhaisajyaguru. He kneels on the lotus seat which grows out of the pool water. It seems to be placed in the air, which shows the wonderful and special view of the Oriental Pure Glass World where the medicine Buddha lives. He looked at Buddha above, and his face is quiet and solemn, the eyebrows and eyes are slender, ears reach to shoulders. The left hand holds the lotus bud, the right hand supports behind on the lotus seat; the posture is graceful, and the expression is attentive.

In terms of dress modeling, this Bodhisattva has a big bun behind his head, and his braided hair flutters. He wears a jewel crown, earrings, neck ornaments, armlets and bracelets. His jewelry is inlaid with blue and green pearls, which used color fading method to show the crystal texture, which is exquisite and beautiful. The Bodhisattva's upper body is naked, wrapped by the heavenly clothes, and the lower body wears a long antariya and the material is soft and flowing. This figure is graceful in posture, vivid in lines, steady in color and elegant in design, which fully demonstrates the typical artistic features of the early Tang Dynasty.

(Text by: Liu Yuanfeng)

图：刘元风 Picture by: Liu Yuanfeng

供养菩萨裙身图案

莫高窟初唐第220窟主室北壁

The pattern on the offering Bodhisattva's antariya

on the north wall of the main chamber in cave 220 dated to the early Tang Dynasty at Mogao Grottoes

供养菩萨裙子的图案为横向分段式，从图像中显露出的部分来看主要分为三段，第一段为方格联珠十字花纹，以红色和白色为底色，边缘为白色线条勾勒的联珠纹和弓字纹。第二段为水波纹，以石青色为底色，用白色线条勾勒段状水波曲线，并穿插墨线联珠，二者错落排列，充满流动感。第三段为菱格十字花纹，以红色为底色，以白线搭建菱形骨架，中间填饰白色四簇花心和石青、石绿色线条。整体图案以细密的几何纹为主，色彩清新雅致。这里选取第二段和第三段图案，并参考常沙娜主编《中国敦煌历代服饰图案》书中类似图案的造型和色彩进行整理绘制。

（文：崔岩）

The pattern on the Bodhisattva antariya is horizontal segmented. From the selected part in the picture we can see it is mainly divided into three sections. The first section is the cross in checkered beads pattern（联珠十字花纹）, with red and white as the background colors, and the trimming is the beads pattern and bow pattern depicted by white lines. The second section is water ripple pattern（水波纹）, with azurite as the background color; the water waves are depicted by white lines and interspersed with small circles depicted by ink lines. The two are arranged in stagger, which seems like flowing. The third section is rhombic cross pattern, with red as the background color and used white lines to build a rhombic frame. The rhombuses are filled by four cluster white flower centers（四簇花心）, azurite and malachite green lines. The overall pattern is mainly composed by fine geometric pattern with fresh and elegant colors. This paper selects the second and the third part, and refers to the shape and color of the similar patterns in the book *Decorative Designs from China Dunhuang Murals* edited by Chang Shana.

（Text by: Cui Yan）

图：王馨、王可 Picture by: Wang Xin, Wang Ke

The Dancer's clothes

on north wall of the main chamber in cave 220 dated to the early Tang Dynasty at Mogao Grottoes

伎乐人服饰

莫高窟初唐第220窟主室北壁

此图为初唐第220窟主室北壁药师经变中乐舞场景中的舞伎之一，表现的是胡旋舞者的背面。她和另一身舞者组成一组，双手握住上下翻飞的长巾，赤足而单脚跳跃于圆毯之上，在药师佛座前灯轮的两侧欢快起舞。

舞者戴杂宝头冠，辫发披散，戴颈饰和手镯，璎珞横飞。上身穿方格纹锦甲半臂，中束腰带，胯间饰以石绿色围腰，并在身后束结为宝珠。下穿卷草纹短裙和褐红色荷叶边阔腿裤，裤子造型与此窟南壁伎乐人的装束相同，手中长巾和腰带前后环飘。画面整体线条和色彩飘逸而富于张力，充分表现出快速旋转、风驰电掣的舞蹈特点，正如唐杜佑《通典》中所述胡旋舞“舞急转如风”的记载，形象地传达出胡旋舞的美妙。胡旋舞是由西域传入、风靡于盛唐的舞蹈，白居易在《胡旋女》中写道：“弦歌一声双袖举，回雪飘飖转蓬舞。左旋右转不知疲，千匝万周无已时。人间物类无可比，奔车轮缓旋风迟。”作为此舞种传播途径中的必经之地——敦煌，在壁画中留下了这一神奇舞蹈的定格瞬间以及伎乐舞者的着装特色。

（文：刘元风）

This is one of the dancers in the music and dance scene of the Bhaisajyaguru tableau in cave 220 dating back to the early Tang Dynasty. It shows the back of the Hu Xuan dancer（胡旋舞者）. She and the other dancer form a pair, holding the long ribbon throwing up and down by both hands, feet bared and jumping by one foot on the round carpet, dancing happily on both sides of the lamp wheel in front of the Buddha's seat.

This dancer wears miscellaneous crown, braided hair, necklaces, bracelets and keyūra. The upper body wears a checkered brocade armour half sleeves, a waist belt in the middle, a malachite green waist wrap around the waist and tied behind as a orb, and a scrolling grass pattern short antariya and brown red ruffled hem wide leg trousers on the lower body（荷叶边阔腿裤）. The trousers are the same as those of the dancers on the south wall of this cave, her long ribbon and belt fluttered around her. The general lines and colors of the image are elegant and full of tension, which fully shows the fast rotation and speed dancing characteristics. As described in the *Tong Dian*（《通典》）of Tang Dynasty, the record of Hu Xuanwu "whirling like wind" vividly conveys the beauty of the dance. Hu Xuanwu is a kind of dance introduced from the Western Regions and was popular in the Tang Dynasty. Bai Juyi wrote in *Hu Xuan Nü*（《胡旋女》）: "In the sound of drum, the sleeves are flying like snowflakes in the air, like the grass flying in the wind. Even the running wheels feel slower than her, even the rapid whirlwind is inferior. She is indefatigable in turning left and right and still turning all the time. Nothing in the world can be compared. She spins faster than the wheels and the wind（弦歌一声双袖举，回雪飘飖转蓬舞。左旋右转不知疲，千匝万周无已时。人间物类无可比，奔车轮缓旋风迟）. Dunhuang, as a inevitable place for the transmission of this kind of dance, preserved a moment of this magical dance in the mural and the dress features of the dancers.

（Text by: Liu Yuanfeng）

图：刘元风　Picture by: Liu Yuanfeng

伎乐人半臂图案

莫高窟初唐第220窟主室北壁

The pattern on the dancer's half sleeves

on the north wall of the main chamber in cave 220 dated to the early Tang Dynasty at Mogao Grottoes

舞者穿贴身窄袖的半臂，从露出的背部来看，图案主体以红色为底色，上绘石绿色方格纹，格子内有白色十字花纹点缀石绿色花心，以及石青色十字花纹，二者错落排列，色彩清雅秀丽。图案整体方正有度，如同铠甲一般，在统一的秩序中不失变化。几何式框架的图案具有规则刚劲的观感，符合胡腾舞所具有的矫健雄迈的特点，因此十分符合舞者身份。绘者参考了常沙娜主编《中国敦煌历代服饰图案》书中相同图案的造型和色彩进行整理绘制。

（文：崔岩）

From this picture we can see that the main body of the pattern is red as background color, with malachite green checkerboard on it. There are white cross patterns dotted with malachite green flower center and azurite cross pattern. The two are arranged at random, and the color is elegant and beautiful. The whole pattern is square and regular, just like armor, and it changes in a unified order. The geometric frame pattern gives a regular and vigorous impression, which is accord with the vigorous and manhood characteristics of Huteng dance（胡腾舞）, so it is very consistent with the identity of the dancer. We referred to Chang Shana's *Decorative Designs from China Dunhuang Murals* for help when doing the drawing.

（Text by: Cui Yan）

图：王可 Picture by: Wang Ke

燃灯菩萨服饰

莫高窟初唐第220窟主室北壁

The lamp lighting Bodhisattva's clothes

on the north wall of the main chamber in cave 220 dated to the early Tang Dynasty at Mogao Grottoes

根据《药师琉璃光如来本愿功德经》所述，供养药师佛的其中一种方法是燃灯："燃四十九灯；造彼如来形象七躯，一一像前各置七灯，一一灯量大如车轮，乃至四十九日，光明不绝。"因此敦煌莫高窟壁画自隋代开始，凡绘制药师经变，均绘出灯轮，侧旁常伴有燃灯菩萨，这身初唐第220窟主室北壁药师经变下部的燃灯菩萨是其中造型最为优美者。她面目安详，双目凝视，蹲于莲座之上，正在虔心燃点手心中的灯盏，递于身旁接应放置灯盏于灯轮上的菩萨。

画面的艺术处理上，线条圆顺而舒朗，色彩明快而雅致。服饰造型上，菩萨戴宝珠头冠，束高髻，长发披肩，戴颈饰和手镯。上身披宽大的天衣，束围腰，下着曳地长裙，博带飘垂。虽然这身燃灯菩萨在整幅经变画中的位置不甚突出，整体造型和装饰亦较为简练，但是绘者通过对人物体态和神情的贴切描绘，以及以石绿色和白色为主的色彩基调，营造出一种宁心静气的审美意境，与周围歌舞声乐的热烈气氛形成了鲜明的对比。

（文：刘元风）

According to the *Bhaisajyaguru-sutra*, one of the ways to worship the Medicine Buddha is to light lamps: "Light 49 lamps, create the image of the Buddha seven bodies, put seven lamps in front of each, every lamp is as big as a wheel, keep the lamps burning for 49 days（燃四十九灯；造彼如来形象七躯，一一像前各置七灯，一一灯量大如车轮，乃至四十九日，光明不绝）." Therefore, since the Sui Dynasty, the murals in Mogao Grottoes have drawn lamp wheels, which are often accompanied by lighting lamp Bodhisattva. Among them, the most beautiful one is the Bodhisattva at the bottom of the north wall of the main chamber in cave 220 dating back to the early Tang Dynasty. His face is serene, his eyes are gazing, and he is squatting on the lotus seat, lighting the lamp in his hand and handing it to the Bodhisattva who is placing the lamp on the lamp wheel.

In the artistic treatment of the picture, the lines are round and smooth, and the colors are bright and elegant. In terms of dress modeling, this Bodhisattva wears jewel crown, high bun; long hairs cover the shoulders, neck ornaments and bracelets. The upper part of the body is covered by broad heavenly robe, waist wrap, and long antariya. Although the position of this lighting lamp Bodhisattva in the whole painting is not prominent, and the overall shape and decoration are relatively simple, but the painter created an aesthetic mood of tranquility through the appropriate depiction of the figure's posture and expression, as well as the color tone is mainly composed by malachite green and white, which forms a sharp contrast with the warm atmosphere of singing and dancing surroundings.

(Text by: Liu Yuanfeng)

图：刘元风 Picture by: Liu Yuanfeng

童子服饰

Children’s clothes

莫高窟初唐第220窟主室南壁

on south wall of the main chamber in cave 220 dated to the early Tang Dynasty at Mogao Grottoes

在初唐典型洞窟莫高窟第220窟的南壁阿弥陀经变壁画上，下部莲花池中绘有两组叠罗汉的童子，他们憨态可掬、天真无邪的形象为壁画描绘的佛国世界盛大恢宏的景象与欢乐祥和的氛围平添了几分人间的乐趣。化生童子的形象在莫高窟北朝洞窟中已经出现，但大多只是简单、程式化的勾勒。唐时期壁画上化生童子的刻画明显更加细腻生动，逼真传神。

莲池西侧的一对叠罗汉的化生童子中，下方的童子穿着一条褐色背带连体瘦裤口长裤，在新疆吐鲁番出土的《双童图》中，也有极为近似的彩色条纹背带长裤造型。上方的童子穿着红色交领半臂，配绿色短裤。半臂是唐前期极为流行的一类服装，当时半臂的形态非常丰富，衣长与领形也富于变化，这件长及胯的半臂与莲池东侧童子穿着的长及臀下的半臂都是当时流行的半臂造型的真实反映。

在莲池东侧的一对叠罗汉的童子中，下方的童子穿着背带上衣、配深褐色短裤。上方的童子穿红绿拼色交领半臂、配褐色条纹长裤。半臂初现于三国时期，在唐前期达到鼎盛，随着中唐以后服装廓型日渐宽大，造型合体的半臂逐渐淡出流行的行列。壁画上化生童子穿着的半臂上红下绿，在腰线处拼合，这一形制与美国芝加哥普利兹克收藏的联珠对鸟纹锦拼色半臂童装极为相似，这几身童子服饰不仅是当时平民世俗生活的折射，也清晰地反映了隋唐童装的造型特征。

（文：李迎军）

In the mural cf Amitabha Sutra tableau on the south wall of cave 220 in Mogao Grottoes dating back to the early Tang Dynasty, there are two groups of children in the lotus pool at the bottom. Their cute and innocent character added a lot of fun to the grand and magnificent scene of some happy and peaceful atmosphere in the Buddhist world depicted in the mural. The images of reincarnated children first appeared in the Northern Dynasties’ caves in Mogao Grottoes, but most of them are simple and stylized. In Tang Dynasty, the depiction of reincarnated children were more alive and vivid.

Among a pair of reincarnated children on the west side of the lotus pond, the boy below is wearing a pair of thin trousers with brown braces. In the unearthed painting from Turpan, Xinjiang, there are two children have very similar colored stripes suspender trousers. The boy above is wearing a red cross collar half sleeves（半臂, just like T-shirt）and green shorts. The half sleeves were very popular clothes in the early Tang Dynasty. At that time, the styles of the half sleeves were very rich, and the length and collar shapes also varied. The half sleeves with length reaching to the waist and the half sleeves reaching beyond the hips wear by the boy on the east side of the lotus pond were the real reflection of the popular half sleeves style at that time.

In the east side of lotus pond, a pair of children are playing a pyramid game. The boy below wears suspender trousers and dark brown shcrts; the boy above wears red and green cross collar half sleeves with brown striped trousers. Half sleeves first appeared in the Three Kingdoms period and reached its peak in the early Tang Dynasty. With the increasing popularity of loose clothes fashion after the Mid Tang Dynasty, the half sleeves with a fitting shape gradually faded out of the fashion ranks. In the mural, the half sleeves of reincarnated children are red on the top and green at the bottom, which meet at the waist line. This shape is very similar to the existing half sleeves children’s clothes collected by Dritzker in Chicago of the United States, which is not only a reflection of the common people’s life at that time, but also clearly reflects the modeling features of the Sui and Tang children’s clothes.

（Text by: Li Yingjun）

图：李迎军　Picture by: Li Yingjun

The heavenly woman's clothes
on the south side of the east gate in cave 220 dated to the early Tang Dynasty at Mogao Grottoes

天女服饰
莫高窟初唐第220窟东壁门南

此为初唐第220窟东壁门南维摩诘经变中的天女。据维摩诘所说经中所载，此天女以法力散花验道，以“花不着身”寓意修行中的分别心；她游戏神通，与舍利弗互换性别，阐释“法无定相”的佛理。天女立于维摩诘的右侧，神情端穆，面庞圆润，额间绘如意形花钿，脸颊一侧绘斜红，弯眉纤细，双目狭长，鼻子高挺，唇型饱满，两侧点红色妆靥，面相雍容，双臂摊开，其左手心置一花朵，右手持麈尾。

从服饰造型上来看，天女头戴双凤宝冠，插如意云头步摇，梳高髻。服装形制为宽缘交领式大袖裙襦，外着石绿色曲折领、红色半袖衫，饰有石绿色翻折荷叶边，点缀石青色四瓣小花图案；内着白色中单，曲领半露；大袖层叠，镶石青色袖缘。腰系石绿色长带，长至脚踝，围红色蔽膝，蔽至胫中，饰有联珠团花纹，缘边用卷草纹装饰。下着白色长裙，盖至脚面，着“凹”字形翘头履。画面中人物丰神异彩，用线俊逸凝练，用色妍丽活泼，彰显风度翩翩的天女形象。

（文：楚艳、苏芮）

This is the heavenly woman in Vimalakirti Sutra tableau in cave 220 dating back to the early Tang Dynasty. According to the Vimalakirti Sutra, this heavenly woman tested people's understanding of Dharma using her magical power to disperse flowers, which implying the differentiate heart in practice of "flowers cannot attached by body"; she also used magical power and exchanged gender with sariputra to demonstrate the Buddhist principle "Dharma without fixed form". She stands on the right side of Vimalakirti, has a solemn expression and a round face; a Ruyi flower pattern on her forehead, red shading on one side of her cheek, slender eyebrows, narrow and long eyes, high nose, full lips, red dimples on both sides, graceful face and outstretched arms. She has a flower in her left hand and a Zhuwei in her right hand.

From the perspective of dress modeling, this lady wears a pair of phoenix crown, inserts Ruyi cloud-shaped end Buyao (hair accessory), and combs high bun. The shape of the dress is a wide margin cross collar large sleeve coat (宽缘交领式大袖裙襦), with a malachite green zigzag collar and red half sleeves shirt, decorated by malachite green folded lotus leaf trimming, scattered azurite four petals flower pattern; the inside is white Zhongdan (中单, a kind of underwear like sweater), the curved collar is half exposed; the large sleeves are stacked and inlaid with azurite trimmings. The waist has a long malachite green belt, which extends to ankles. The legs are covered by red Bixi (蔽膝, a piece of cloth covering from waist to lower legs) reach to lower legs. It is decorated with beads pattern, and the trimming decorated with scrolling grass pattern. The lower body wears the white long antariya covering to the feet and she wears "concave" shaped shoes. The figure in the picture has bright and lively colors, concise lines, showing the graceful image of the heavenly woman.

(Text by: Chu Yan, Su Rui)

图：楚艳 Picture by: Chu Yan

天女蔽膝图案

莫高窟初唐第220窟东壁门南

The pattern on the heavenly woman's Bixi

on south side of the east gate in cave 220 dated to the early Tang Dynasty at Mogao Grottoes

敦煌莫高窟第220窟东壁门南所绘天女的半袖衫和蔽膝上均装饰有丰富的服饰图案，这里选取其蔽膝图案进行整理绘制。蔽膝的主体图案以大面积的土红色为底色，上绘排列有序的联珠团花纹。图案以白色实心点为花心，外绘石青色圆圈，外周为一圈白色联珠，联珠以放射性白线与花心连接在一起，状如花蕊。宾花为石青色花心的菱形联珠纹，也呈放射状，穿插在圆形轮廓的联珠团花纹之间。蔽膝镶着较宽的白地缘边，上以土红色线勾勒出二方连续式卷草纹。整体图案整体清新秀丽，灵动雅致。

（文：崔岩）

There are rich patterns on the half sleeves and Bixi (a piece of cloth covering legs to knees) of the heavenly woman painted on the south side of the east gate in cave 220 of Mogao Grottoes. Here, we choose the pattern of Bixi to analyse. The main pattern of the Bixi is a large area of earth red as the background color, with orderly arranged beads pattern painted on it. The pattern takes the white solid dots as the flower center, it has azurite circles on the outside and a circle of white dots on the periphery. The flower is a rhombus shaped beads pattern with a azurite flower center, which is also radial and interspersed between the round shaped patterns. The Bixi is inlaid with a wide white border, and the two in a group repeated scrolling grass pattern outlined by red lines. The overall design is fresh and beautiful, flexible and elegant.

(Text by: Cui Yan)

图：常青　Picture by: Chang Qing

二身异国王子服饰

莫高窟初唐第220窟东壁门南

Two foreign princes' clothes

on the south side of the east gate in cave 220 dated to the early Tang Dynasty at Mogao Grottoes

敦煌莫高窟初唐第220窟东壁门南维摩诘经变中绘有各国王子听法图，对应的是《维摩诘所说经·方便品第二》中关于维摩诘"以其疾故，国王大臣、长者居士、婆罗门等，及诸王子，并余官属，无数千人，皆往问疾"这一故事情节，这里选取其中的二身异国王子进行服饰效果图整理绘制。

二身王子高鼻深目，神情专注，身体侧向前方，双手揣在袖中，平举在胸前，似在安静地聆听文殊菩萨和维摩诘居士辩论说法。左侧这身王子，头戴花锦浑脱帽，两耳垂耳珰，身着石绿色的圆领镶边锦袍，外披毡袍，足穿长筒乌靴。这种浑脱帽是用织锦制作而成，并有细毛毡镶边。帽的顶部隆起，以白色为底色，主体装饰有石绿色和石青色的菱形、如意形及团花纹样，华丽异常。身着的锦袍边缘有褐色底的二方连续式团花纹缘饰，组织细密，纹彩兼备，有的还加上金线装饰，更突出了其身份的尊贵。其形象可能为波斯人，或是来自中东地区的使者。右侧的王子头部较为模糊，同样身着红色圆领花边锦袍，领、袖和衣襟处为二方连续式波浪纹，锦袍下摆镶宽缘边，装饰着四瓣团花纹。袍长至脚踝处，脚着红色皮靴，其服装形制和装饰风格都充满着异域色彩。

（文：楚艳、王子怡）

There is a painting of princes listening to Dharma on the south side of the east gate in cave 220 dating back to the early Tang Dynasty in Dunhuang Mogao Grottoes, which is a part of Vimalakirti Sutra tableau. It corresponds to the the story of "King, ministers, elders, Brahmins, princes, and other officials, and thousands of people, all go to inquire about the disease" in *Vimalakirti Sutra the skillful means chapter, the second*. This paper selects two foreign princes to analyse their clothes.

The two princes, with high noses and deep eyes, concentrated expressions, leaning forward with hands in their sleeves and hold them in front of the chests. They seemed to be listening to the debate quietly between Manjusri Bodhisattva and Vimalakirti. The prince on the left is wearing a Huntuo hat（浑脱帽）, earring, a malachite green coat with round collar, covered with felt robe, feet in long black boots. This kind of hat is made of brocade with fine felt edging. The top of the hat is raised, with white as the background color. The main body is decorated with rhombus pattern, Ruyi pattern and flower medallion pattern in malachite green and azurite color, which is gorgeous and unusual. On the trimming of the robe, there are two in a group repeated flower medallion pattern fringes on the brown bottom; both patterns and colors are well-organized. Some of them are decorated with gold thread, which highlights their dignity. This image may be a Persian or a emissary from the Middle East. On the right, the prince's head is fuzzy, and he is also wearing a red round collar lace trimming robe. The collar, sleeve and lapel of the robe have two in a group repeated wavy patterns. The hem of the robe is inlaid with wide trimming and decorated with four petals flower medallion pattern. The robe reaches to the ankles and his feet in red leather boots. The clothes design and decorative style are full of exotic styles.

（Text by: Chu Yan, Wang Ziyi）

图：楚艳 Picture by: Chu Yan

弟子像袈裟内里、裙身图案

莫高窟初唐第220窟主室西壁龛内北侧

The pattern on disciple's kasaya lining and robe

Mogao grottoes cave 220 dated to the early Tang Dynasty, the north side of the west niche in the main chamber

第220窟主室西壁龛内壁画与龛内彩塑相互呼应，绘制弟子像、菩萨、背光等内容，现将北侧弟子像的服饰图案进行整理绘制。弟子外披袒右浅灰色田相纹袈裟，内披石绿色袈裟，在袈裟的翻折处露出内里。图案底色为浅石绿色，以墨线勾勒几何图案为主花，花型为两道半圆弧线呈相背对称状，在上下左右点缀黑色实心圆点，造型简练，清秀雅致。弟子所着长裙为浅绛色，白色花纹为主花。花纹为十字形骨架，装饰四出心形叶片，空隙以白色圆点填充。图案整体呈四方连续形式，花地分明，清地排列，错落有序。

（文：王馨）

The murals in the west niche of the main chamber in cave 220 echo with the painted sculptures in the niche. The image of disciples, Bodhisattvas and backlights are painted. Now, the clothes patterns of the disciples' image on the north side are analysed. The disciple wears right shoulder bared light gray Kasaya with field pattern on the outside and malachite green Sankaksika inside. The inside of the kasaya was revealed at the folds of the robe. The background color of the pattern is malachite green, which used ink lines to outline the geometric pattern. The flower pattern is two semicircular arcs, which are symmetrical to each other; black solid dots are scattered on the top, bottom, left and right. The shape is simple and elegant. The long dhoti worn by the disciple is light crimson with white pattern as the main flower. The pattern has a cross shaped frame, decorated with four heart-shaped leaves, and the gap is filled with white dots. The whole pattern is repeat in four as a group, with distinct flower shape, clearly and orderly arranged.

(Text by: Wang Xin)

图：常青 Picture by: Chang Qing

The emperor clothes

on the north side of the east gate in cave 220 dated to the early Tang Dynasty at Mogao Grottoes

帝王听法图服饰

莫高窟初唐第220窟东壁门

敦煌莫高窟初唐第220窟东壁门两侧绘有维摩诘经变，画面下部分别为帝王听法图和各国王子听法图，对应的是《维摩诘所说经·方便品第二》中维摩诘"以其方便，现身有疾"，国王大臣、诸王子等众人前来问疾的情节。这里选取的是东壁门北帝王听法图中的帝王像进行服饰整理绘制。

帝王头戴冕冠（不知为何只有六旒，而非十二旒），着上玄下纁的冕服，内为白色曲领中单，以菱格纹大带束绛色蔽膝，着云纹笏头履。从图像中可以辨别出十二章纹中的七种：在上衣的两肩各画一圆圈，一侧为鸟形，为金乌表示太阳，一侧画玉兔，表示月亮；在袖身上可以看到山岳和龙纹；绛色蔽膝上为火纹；袖缘上若干白色小点组合形成的小花形为粉米；在袖口有类似亚字形，为黻纹。后面的群臣为文官，戴黑介帻，穿具服。

整体而言，帝王听法图不仅描绘了唐代帝王与群臣的威仪之姿，为我们提供了丰富的服饰资料，还展示了当时佛教的兴盛与人们对佛教的尊崇。

（文：楚艳、王子怡）

On both sides of the east gate of cave 220 dating back to the early Tang Dynasty in Dunhuang Mogao Grottoes have the Vimalakirti Sutra tableau; there is a portrait of an emperor listening to Dharma and a group of princes listening to Dharma at lower part of the wall, which corresponds to the plot of *Vimalakirti Sutra· skillful means chapter, the second* that Vimalakirti "appears sick", and the kings, ministers, princes and other people come to visit. Here we choose the image of the emperor listening to Dharma in the picture on the north side of the east gate to analyze the clothes.

The emperor wears a Mian crown（冕冠, royal crown, don't know why there are only six strings instead of twelve strings）on his head, black robe on the upper body and yellow red clothes on the lower body; a white curved collar Zhongdan（曲领中单）, using a large rhombus pattern belt ties the red Bixi, which covered his knees with crimson color; he wears cloud pattern square shaped head shoes. We can identify seven symbols out of Twelve Symbols on the clothes, for example there are two circles on each shoulder, one side has a bird in the circle means Jinwu（金乌, a black crow）which representing the sun, the other side has a rabbit means Yutu（玉兔）which representing the moon. On sleeves have mountain pattern, dragon pattern and fire pattern on Bixi; some white dots on the cuffs representing flour and rice, 亚shaped（黻纹）pattern on the cuffs. Behind the emperor are officials for civil servants, wearing black Jieze（介帻, a kind of hat wear by officials）Jufu（a kind of uniform wears by officials）.

In general, the emperor listening to Dharma painting not only depicts the dignified posture of the emperor and his ministers in the Tang Dynasty which provide us abundant clothes information, but also shows the prosperity of Buddhism and the respect to Buddhism at that time.

（Text by: Chu Yan, Wang Ziyi）

图：楚艳 Picture by: Chu Yan

侍从服饰

莫高窟初唐第220窟东壁门北

The servant's clothes on north side of the east gate in cave 220 dated to the early Tang Dynasty at Mogao Grottoes

主室东壁的维摩诘经变中，位于门北侧的为帝王掌扇的侍从形象曾经被多位敦煌研究院的画家临绘过，但由于这个人物位于接近地面的门边，所以残破损毁较严重，因此虽然有多幅临绘作品可以参照，但他的服饰仍有几处“特殊”之处有待继续研究。

掌扇侍从身着朱色大袖衫，为了便于活动而将宽大的袖口打结，露出了内搭的合体绿色小袖。大袖衫外罩朱色裲裆，束白腰带，铊尾上翻。据《谈苑》记载，唐高祖曾下令禁止“反插垂头”，因此自初唐以后，如图所示的反插铊尾的形态就消失了。下着白色缚裤，由于裤腿之下已残缺不清，所以无法准确分辨脚上的穿着。同时期的袴褶下多搭配鞋靴，因此在绘画整理时，参照之前临摹壁画上的形态画成了鞋。

袴褶自魏晋至隋唐始终是仪仗卫队的标准服饰，以平巾帻、大袖衫、裲裆、袴褶为标配。第220窟这位掌扇侍从形象的帽子只有线描造型，与其他侍从均戴平巾帻相比，这一侍卫的帽子似是尚未画完，且从现有的线描造型看也是状如翻边毡帽而非平巾帻。之前壁画的临摹资料也大多延续了毡帽的形态，因此在绘画整理时暂且保留了与目前举目可见的线描造型相一致的帽形。

（文：李迎军）

In the Vimalakirti Sutra tableau on the east wall of the main chamber, the image of the servant holding a fan on the north side of the gate was once copied by several Dunhuang Academy painters. However, because this figure is on the side of the door near to the ground, so it is seriously damaged. Therefore, although there are many copies to refer to, several parts of his clothes need to be further studied.

The servant is wearing a red wide sleeves coat. In order to do job conveniently, the wide cuffs are tied, revealing the green sleeves shirt. The big sleeves coat is covered by red Liangdang (a kind of vest), white belt and the belt's end turning up. According to *Tanyuan*《谈苑》, Emperor Gaozu of the Tang Dynasty once ordered the prohibition of "people can not put the end of the belt upright". Therefore, since the beginning of the Tang Dynasty, the upright belt end as shown in the figure has disappeared. The lower body wears the white tied pants (白色缚裤); because the feet are incomplete, we don't know what he wears on the feet. At that time flat shoes and boots were often used to match the pants. Therefore, flats were painted according to the shape of the mural copied before.

From the Wei and Jin Dynasties to the Sui and Tang Dynasties, Kuzhe (袴褶) has always been the standard clothes of honor guards, with Pingjinze (平巾帻, a kind of hat,originally like a kerchief), large sleeve coat (大袖衫), Liangdang (裲裆, the vest) and pants as the standard. In cave 220, the hat of the servant only has line drawing. Compared with other attendants who wear Pingjinze, the hat of this servant seems to have not been finished, and according to the existing line drawing modeling, it looks like a flanging felt (翻边毡帽) hat rather than a Pingjinze. Most of the copying materials of this mural before also used the form of felt hat, so the outline of the hat which still visible part was used to finish the hat shape here.

(Text by: Li Yingjun)

图：李迎军　Picture by: Li Yingjun

弟子像袈裟图案
莫高窟初唐第220窟主室西壁龛内南侧

The pattern on disciples' kasaya
on the south side of the west niche in the main chamber of cave 220 dated to the early Tang Dynasty at Mogao Grottoes

图案取自敦煌莫高窟第220窟主室西壁龛内南侧弟子像的袈裟图案。图案以浅土黄为底色，在田相格内绘石青色山丘，山峦上勾勒石绿色渲染山野间清秀的景致，点缀些许赭石色使图案整体更加沉稳有力。

这种饰以山水林木图案的袈裟被称为“山水衲”，这是从初唐出现并沿袭的袈裟样式，晚唐第17窟洪辩塑像仍着此类袈裟。宋人《四分律行事钞资持记》中记载说：“然此衣竝世人所弃零碎布帛收拾斗缀以为法衣，欲令节俭少欲省事，一纳之外更无余物。今时禅众多作纳衫而非法服，剪裁缯綵，刺缀花纹，号山水衲，价直数千，更乃各斗新奇，全乖节俭。”从记载中可以得知，这种山水衲不同于法衣，它不仅有裁剪，而且运用高档面料，此外还加饰了多种装饰工艺表现山水纹，是一种十分讲究的僧侣服装。

（文：王馨、崔岩）

This design is taken from the kasaya worn by the disciple in the south side of the west niche of cave 220 in Dunhuang Mogao Grottoes. The background color is light earth yellow, and the malachite green hills are painted in field pattern. The mountains are outlined in malachite green, rendering the beautiful scenery among the mountains, and dotted with some ochre color, which makes the overall design more stable and powerful.

This kind of kasaya decorated with the pattern of mountains, rivers and trees is so called “Shanshuina”（山水衲）. This kind of kasaya appeared in the early Tang Dynasty. According to the Song Dynasty's *Sifenlvxingshichaozichiji*（《四分律行事钞资持记》, notes for Dharmagupta Vinaya）, it is said: “Kasaya should be made by pieces of cloth and silk discarded by people, it was intended to make people frugal and simple, to have nothing more. Today's clothes are made by expensive materials, decorated by beautiful patterns, named Shanshuina. It is very expensive and everyone is competing for new style, definitely not frugal” We can learn from this that Shanshuina is different from the original kasaya, it is not only well tailored but also uses high-grade fabrics. Meanwhile it has a variety of decorative techniques to show the landscape patterns. This is a kind of very exquisite monk clothes.

（Text by: Wang Xin, Cui Yan）

图：王可　Picture by: Wang Ke

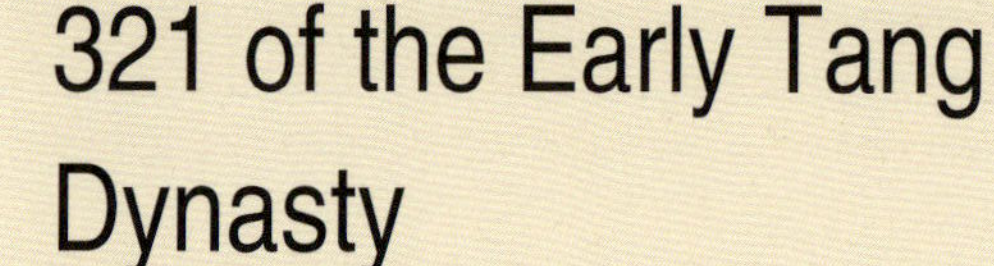

敦煌莫高窟初唐

第321窟

Dunhuang Mogao Grottoes Cave

321 of the Early Tang Dynasty

第321窟的前室已坍毁，为中小型覆斗顶窟，是初唐时期的代表窟之一。主室窟顶为团花卷草纹藻井，四披画千佛。西壁开凿平顶敞口龛，彩塑一佛、二弟子、二菩萨、二力士，经清代重修。龛顶绘赴会佛三铺及天宫天人、飞天等，两侧绘飞天、菩提树及头光等。龛沿绘卷草纹边饰，龛外上部绘七佛，两侧绘千佛，力士台上存清塑天兽各一身，南侧台下存初唐绘菩萨两身。主室南壁绘十轮经变，宣传地藏菩萨信仰，是敦煌壁画中不可多见的题材。北壁绘阿弥陀经变，碧波蓝空，天水相接，以巍峨的楼宇殿阁为背景，围绕主尊阿弥陀佛穿插安置众多的菩萨、飞天、伎乐、动物等，表现了西方净土庄严清净的美妙景象。主室东壁门上绘倚坐佛说法图一铺，两侧趺坐佛说法图各一铺，门南绘立佛与菩萨，门北绘十一面观音经变。主室北壁和东壁门下部均为五代所绘男供养人像，较为模糊。前室仅存西壁，残存地狱变等初唐画迹多已模糊。甬道为盝形顶，有五代重绘壁画，多已残毁。

此窟壁画在绘画题材上体现出净土宗和密宗融合的佛教思想，用色沉稳大胆，善用土红、石青、石绿等色彩退晕进行对比表现。主室西壁龛顶和北壁的天空均以大面积的蓝色铺地，表现澄明通透的天界，以此衬托莲池海会中飞天翱翔、楼阁耸立、云气缭绕、不鼓自鸣的祥和场面。人物造型丰腴壮硕，体现了唐代强劲有力的审美风格，服饰风格和表现手法较为多元，注重线面结合。壁画中的图案丰富多彩，云纹、花砖、经幢、华盖、植物等各类图案色彩明艳精炼，具有明显的平面化和装饰化的倾向，体现了初唐构图渐趋繁复、设色明快雅致、注重画面细节的艺术处理。

南壁十轮经变中部和下部有大量的菩萨、弟子和世俗人物服饰表现。世俗人物中出现多个有中原服饰特点的形象，色彩淡雅。中部菩萨服饰颜色饱和，其中也出现了中原服饰，如圆领大袖袍服。东壁菩萨服饰以披帛和裤为主，搭配长披巾和围腰，注重表现面料透明的质感。北壁服饰色彩已经不能分辨清楚，成组的菩萨以长裙装束为主，线条非常优美，整体造型十分饱满。

（文：崔岩）

The front chamber of cave 321 has collapsed. It is a small and medium-sized truncated pyramidal ceiling cave, one of the representative caves in the early Tang Dynasty. At the top of the main chamber is an caisson well with scrolling grass pattern and flower medallion pattern, the four slopes covered by Thousand Buddha motif. On the west wall, there is a flat top open niche is excavated, with colored sculptures of one Buddha, two disciples, two Bodhisattvas and two protectors, which were remade in the Qing Dynasty. The top of the niche is painted with three Buddhas going to preaching, celestial beings, flying Apsarases, etc., and the two sides are painted with flying Apsarases, Bodhi trees and head lights, etc. There are seven Buddhas on the upper part of the niche and Thousand Buddha motif on both sides. On the protector's platform, there are two painted sculpture guardian animals made in the Qing Dynasty, two bodies of Bodhisattvas painted in the early Tang Dynasty below the south platform. The south wall of the main chamber is painted with Dasacakra Ksitigarbha Sutra tableau to publicize the belief of Ksitigarbha Bodhisattva, which is a rare subject in Dunhuang murals. The north wall is painted with Amitabha sutra tableau, green water with waves connected to the blue sky. Taking the palace buildings as the background, many Bodhisattvas, flying Apsarases, performers, animals and so on are placed around the main Amitabha Buddha, which shows the solemn and beautiful scene of the pure land in the West. On the east wall of the main chamber, above the door there is a picture of sitting Buddha with feet on ground, two sitting Buddhas on both sides with legs folded. On the south side of the door have a standing Buddha and a Bodhisattva, and at the north side of the door has a picture of eleven headed Avalokitesvara. The lower part of the north wall and the east wall of the main chamber are the portraits of male donors painted in the Five Dynasties, most of which faded. The front chamber only the west wall remains, most of the paintings in the early Tang Dynasty are blurred. The corridor's ceiling is flat, with murals redrawn in the Five Dynasties; most of them have been destroyed.

The murals in this cave embody the Buddhism thought the fusion of Pure Land sect and esoteric sect. The colors are calm and bold; they used earth read, azurite, malachite green and some other colors. The sky on the top of the west niche and the north wall of the main chamber is covered by large areas of blue color, which is trying to show the clear and transparent Heaven; the blue sky complements the peaceful scene of flying Apsarases, towering pavilions, cloud shrouded and self playing instruments. The figures are plump and strong, reflecting the aesthetic taste of the Tang Dynasty. The clothes style and expression methods are diversified, and the combination of line and plane are emphasized. The patterns in the murals are rich and colorful. The patterns' color, such as cloud, tile, dhvaja, canopy, plant, are bright and refined. They have an obvious tendency of planarization and decoration, which reflects the artistic tendency of more complex in composition, bright and elegant in colors, and paying more attention to details in the early Tang Dynasty.

There are large number of Bodhisattvas, disciples and secular characters in the middle and lower parts of Dasacakra Ksitigarbha Sutra tableau on the south wall; some secular figures have the characteristics of Central Plains clothes style, the colors are plain and simple. The Bodhisattvas' clothes in the middle are saturated, and also have Central Plains style, such as round collar and big sleeves robes. The east wall Bodhisattvas' clothes are mainly silk wraps and pants, with long shawls and waist wraps, which are more focusing on the transparent texture of the fabric. The north wall clothes colors are unable to distinguish clearly; groups of Bodhisattvas mainly dress in long antariyas; lines are very beautiful; the overall shape is very plump.

(Text by: Cui Yan)

天女服饰

莫高窟初唐第321窟主室南壁

The Celestial woman's clothes

on south wall of the main chamber in cave 321 dated to the early Tang Dynasty at Mogao Grottoes

这是初唐第321窟主室南壁十轮经变佛陀左下方的三位天女之一。按《大方广十轮经·诸天女问四大品第二》所述，佛陀演说解释地藏菩萨的种种德行，当时有功德天女、功德乐天女、妙音声天女及众多天人、鬼神询问佛陀关于四大根本初、中、后相的起灭因缘。这里所绘应是经文所述的天女之一。

天女面相丰润细腻，朱唇翠眉，昂首行进，双手合掌，默思静虑，侧面转身，面向佛座。服饰造型上，天女头束宝珠发冠，身穿石绿色半袖长衫，下摆处的宽缘边有图案装饰。内套大袖襦，袖口有石绿色缘边。下着长裙，亦饰有石绿色缘边，束腰带，穿花头履。领部为云肩式翻折领，缀有璎珞，半袖处有侈口的荷叶边装饰，长襦下杂裾垂髾，华袿飘飘，加上甩动的头冠束带和璎珞珠串，无不表现出一种凌空平步、凭天翱翔的审美感受。画面艺术处理上，用线刚劲而潇洒。在色彩方面，除了保留原壁画中的石绿色之外，还将已经变色的服饰部分进行了分析处理，力求沉静而典雅。

（文：刘元风）

This is one of the three heavenly women at the bottom left of the Dasacakra Ksitigarbha Sutra tableau in the south wall of the main chamber of cave 321 dating back to the early Tang Dynasty. According to *Dasacakra Ksitigarbha Sutra, the second part of the celestial women asking chapter*（《大方广十轮经·诸天女问四大品第二》）, Buddha explained the various virtues of the Bodhisattva Ksitigarbha. At that time, there were virtue celestial woman（功德天女）, virtious happy celestial woman（功德乐天女）, beautiful voice celestial woman（妙音声天女）, and many other heavenly beings, ghosts and gods who asked the Buddha about the causes and conditions of the rise, develop and cease stages of the four noble truth. The painting here should be one of the heavenly women mentioned in the scripture.

This celestial woman's face is rich and delicate, her lips are red and eyebrows are green; the head is proud, palms hold together, quietly thinking. She turns a little bit and faces the Buddha seat. In terms of dress, this celestial woman wears a jewel crown on her head, a half sleeves shirt in malachite green, and decoration on the wide trimming. She wears large sleeves dress inside with malachite green trimming. The lower body wears a long antariya with malachite green trimming, a belt and flowery shoes. Around the neck is a cloud folded collar（云肩式翻折领）, decorated with keyūra; the half sleeves is decorated with lotus frills. Under the long antariya, the frills hanging and ribbons floating, match with the swinging headband and the string of keyūra; all of them show an aesthetic feeling of flying in the sky. In the artistic treatment of the picture, the lines are vigorous and natural, besides the malachite green in the original murals; the oxidation color of the clothes are analyzed and recovered, aiming for quietness and elegance.

（Text by: Liu Yuanfeng）

图：刘元风 Picture by: Liu Yuanfeng

供养菩萨服饰

莫高窟初唐第321窟主室东壁门北侧

The offering Bodhisattva's clothes

on the north side of the east gate in the main chamber of cave 321 dated to the early Tang Dynasty at Mogao Grottoes

这是初唐第321窟主室东壁门上北侧说法图中的供养菩萨，她面相丰腴，双手捧持长柄香炉，侧身站立在覆瓣莲花座上，面向中央正在说法的佛陀。

服饰特征上，菩萨头梳高髻，辫发披肩，戴镶嵌着宝珠的日月冠。日月冠是一种在宝珠上饰以新月、月中托日的菩萨冠式之一，在敦煌早期壁画中非常流行，至唐代菩萨画像中仍能见到。日月冠是来自波斯萨珊王朝的王冠装饰，这是佛教造像在流传过程中受到沿途地区装饰艺术影响的一个典型例子。菩萨颈部佩戴项链，腕间戴镯，均镶嵌光彩夺目的宝珠，上臂饰臂钏。上身斜披土红色披帛，束腰带，裹石绿色围腰。下着透明纱质的土红色印花裤，为阔腿收口型，且裤子的前片薄如蝉翼，使其曼妙的身姿显露无遗，颇具时尚感，此类造型和质感的裤子在初唐洞窟供养菩萨的服饰表现中较为常见，应受到东南亚地区民族传统服饰的影响。披帛、围腰和裤子上均装饰有粉绿心或土红心的白色五瓣小花，清丽雅致，华贵之气与佛法氛围交相辉映。

（文：刘元风）

This Bodhisattva is located at the north upper part of the east wall in the main chamber of cave 321 dating back to the early Tang Dynasty. She has a plump face, holding a long handle incense burner by both hands, and standing on the flat lotus seat, facing to the Buddha in the center.

In terms of dress features, this Bodhisattva wears a high bun; braided hair cover the shoulders, and she wears a sun moon crown inlaid with jewels. Sun moon crown is one of the forms of Bodhisattva crown decorated with a crescent moon and a floating sun in the center. It was very popular in early Dunhuang murals and can still be seen in Bodhisattva portraits of the Tang Dynasty. The sun moon crown is a crown decoration from the Sassanian Empire in Persia, which is a typical example of Buddhist images influenced by the decorative arts along the way in the process of spreading. The Bodhisattva wears necklace on the neck, and bracelets on the wrists, all of which are inlaid with dazzling pearls, and their upper arms are decorated with armlets. The upper body is obliquely covered with earth red uttariya, belt and malachite green waist wrap. The lower body wears earth red printed pants with transparent texture, wide legs necking type, and the front side of the pants as thin as cicada wings, which shows the graceful posture and very fashionable. Such pants with this shape and texture are common in offering Bodhisattvas' clothes to caves in the early Tang Dynasty, and should be influenced by the traditional clothes from southeast Asia. The white five petaled flowers with pink green center or earth red center are decorated on the uttariya, waist wrap and pants, which are clean and elegant; the noble atmosphere and Buddhist atmosphere complement each other.

(Text by: Liu Yuanfeng)

图：刘元风　Picture by: Liu Yuanfeng

供养菩萨服饰
莫高窟初唐第321窟主室东壁门北侧

The offering Bodhisattva's clothes on the north side of the east gate in the main chamber of cave 321 dated to the early Tang Dynasty at Mogao Grottoes

这里所选取的是初唐第321窟主室东壁门北侧十一面观音经变中的供养菩萨，其面相丰润温婉，垂目静思，左臂下垂，用手掀起衣衫一角，右手在胸前轻拈珠串，正面站立在仰瓣莲花座上。菩萨身躯呈S形造型，姿态优雅，玉树临风，仪态万千。

服饰造型上，菩萨头戴宝珠花冠，颈部饰璎珞，并从两肩垂下石绿色和浅土红色的飘带。上身袒裸，披轻薄透明的敞衣和天衣，衣衫用带子系结，缘边有半破式二方连续结构的卷叶花状图案装饰。图案以土红色为底色，单位图案纵向排列，并以石绿色、石青色的退晕法表现花叶的色彩渐变，以白色线条勾勒边缘。此菩萨的服饰图案与其周围的单独图案在造型和色彩上相互呼应，是疏丽清朗的典型初唐装饰风格。菩萨束石青色腰带，裹褐色镶边的白色围腰，围腰下垂出两条彩色边饰，图案与敞衣缘边相同。下着白色的曳地长裙，有石青色镶边。整体画面处理，线条流畅而飘逸，色彩淡雅而柔和，突出表现了丝织品轻柔而透明的品质。

（文：刘元风）

Here is the offering Bodhisattva in the eleven headed Avalokitesvara Sutra tableau on the north side of the east gate in the main chamber of cave 321 dating back to the early Tang Dynasty. His face is plump and gentle, eyes looking down and contemplating quietly. His left arm is downwards, lifting the corner of his clothes, the right hand pinching the bead in front of his chest; he stands on the inverted petals lotus seat. This Bodhisattva's body is S-shaped, elegant and graceful.

In terms of clothes, the Bodhisattva wears a jeweled crown on his head, keyūra on the neck, and malachite green and light earth red ribbons hanging on his shoulders. His upper body is naked, covered with light and transparent open clothes and celestial clothes. The clothes is tied by lace, and the trimmings are decorated with half flower medallion two in a group repeated structure curled leaf flower pattern. The pattern is arranged vertically with earth red as the background color, and the color gradients of flowers and leaves are displayed by the method of color-gradation technique by malachite green and azurite, and the trimmings are outlined by white lines. This Bodhisattva's dress pattern and its surrounding individual patterns echo with each other in shape and color, which is a typical early Tang Dynasty decorative style. This Bodhisattva wears a azurite belt and a white waist wrap with brown trimming. The waist wrap has two colored fringes hanging, and the pattern as same as the open clothes on the upper body. The lower body wears a long white antariya with azurite trimming. The overall picture is smooth and elegant in lines, soft and simple in colors, highlighting the soft and transparent quality of silk fabrics.

(Text by: Liu Yuanfeng)

图：刘元风 Picture by: Liu Yuanfeng

敦煌莫高窟初唐

第322窟

Dunhuang Mogao Grottoes Cave 322 of the Early Tang Dynasty

莫高窟第322窟始建于初唐时期，五代重修，为覆斗顶小型窟。主室正面开双层龛，内塑一佛、二弟子、二菩萨、二天王，龛内绘弟子像等，外层龛壁分别绘夜半逾城、乘象入胎等佛传故事。西壁龛外两侧各绘一身供养菩萨和维摩诘经变。窟顶绘缠枝葡萄纹飞天藻井，四披绘千佛。南壁和北壁均绘千佛，中央绘弥勒说法图和阿弥陀经变各一铺。东壁门上绘说法图三铺，门南侧绘药师说法图，北侧为一立身菩萨和四跏趺坐佛。甬道盝顶中央为五代绘跏趺坐佛一铺，南北壁绘跏趺坐佛。

此窟以塑像完整及其形象特殊而著名，壁画以土红涂地，属于由隋代风格向初唐风格转变的典型洞窟。主室西壁龛内塑像为离壁圆塑，增强了一铺塑像的整体感，虽已改变隋代彩塑背贴龛壁的贴塑形式，但塑像形体略显单薄，还带有隋代彩塑的造型影响。类似胡人的天王形象尤其突出，显示出塑像受到西域民族面貌特征的影响。说法图中的菩萨身姿已渐秀丽，线条圆润有力，整体色调呈现出富丽而淡雅的初唐风尚。

西龛内彩塑服饰表现清晰，结构和色彩基本完整，装饰纹样大体可见。虽说风格是处于由隋向初唐的过渡阶段，但是从彩塑服装的边饰上来看已经是典型的唐代的风格特征了。西龛后壁弟子服饰层次分明、结构完整，可以看得出不同的穿搭方式。南北两壁说法图中，佛菩萨服饰结构非常清晰，可以看到纺织品纹样的细腻表现，以及唐代典型的团花和半团花边饰的特征，尤其是北壁菩萨披帛中的扎经染，细节丰富完整，与第57窟服饰纹样有异曲同工之妙。菩萨长裙如倒置的花冠一样优美，并以璎珞缠绕出更多丰富的结构性变化。东壁的佛菩萨服饰相对较为简单明了，以大色块和线描绘为主，但是也有如同南北两壁一样的半透明质感的面料表现。

（文：崔岩）

DUNHUANG MOGAO GROTTOES
CAVE 322

Cave 322 of Mogao Grottoes is a small truncated pyramidal ceiling cave built in the early Tang Dynasty and restored in the Five Dynasties. The west wall of the main chamber has a double-layer niche with one Buddha, two disciples, two Bodhisattvas and two heavenly kings painted sculptures inside as well as painted disciples inside the niche. The outer niche walls are painted stories about the Great Departure and the Great Conception. On both sides of the west niche are painted a body of Bodhisattva and Vimalakirti Sutra tableau. The top of the cave is painted with entwined grape branches pattern and flying Apsaras caisson, the four slopes covered by Thousand Buddha motif. Both the south wall and the north wall are covered by Thousand Buddha motif and the central parts are painted with Maitreya Buddha preaching scene and Amitabha Sutra tableau. Above the east gate are three pictures of preaching Buddhas, on the south side of the gate is Bhaisajyaguru preaching scene, and on the north side are a standing Bodhisattva and four sitting Buddhas. The top of the corridor in the center is a picture of sitting Buddha painted in the Five Dynasties, and sitting Buddha painted on the north and south walls.

This cave is famous for its complete sculptures and their special characteristics. The murals are painted by earth red as back ground, a typical cave that has changed from the Sui Dynasty style to the early Tang Dynasty style. The sculptures in the west niche of the main chamber are independent and detached from the wall, which enhances the overall feeling of the group sculptures. Although the back of painted sculptures attaching to the niche wall of the Sui Dynasty style has changed, the shape of the sculptures are slightly thin and have the influence of the Sui Dynasty style. The heavenly kings sculptures look similar to minority people are particularly prominent, which shows that the sculptures are influenced by the features of the western regions. The Bodhisattva's posture in the preaching scene is gradually becoming soft and beautiful, the lines are mellow and powerful, and the overall tone shows the rich and elegant fashion of the early Tang Dynasty.

In the west niche, the painted sculptures' clothes are clear, the structure and color are basically complete, and the decorative patterns are generally visible. Although the style was in the transition stage from Sui Dynasty to early Tang Dynasty, the clothes trimming's decoration is typical style of the Tang Dynasty. The clothes of the disciples on the back wall of the west niche have distinct layers and complete structures, so we can see different ways of dressing. The structure of Buddha and Bodhisattva's clothes are very clear in the pictures on the two walls, we can see the delicate performance of textile patterns, as well as the characteristics of typical round flower and half round flower lace decorations in the Tang Dynasty. In particular, the tie-dyeing in the Bodhisattva's uttariya on the north wall has rich and complete details, which is similar to the clothes patterns in cave 57. The Bodhisattva's long antariya as beautiful as an inverted corolla, and more abundant structural changes are twined with its keyūra. The clothes of Buddhas and Bodhisattvas on the east wall are relatively simple and clear. They are depicted with large color blocks and lines, but they also have the same translucent texture as the north and south walls.

(Text by: Cui Yan)

The offering Bodhisattva's clothes

on the south side of the east gate in the main chamber of cave 322 dated to the early Tang Dynasty at Mogao Grottoes

供养菩萨服饰

莫高窟初唐第322窟主室东壁门南

初唐第322窟主室东壁门南绘有一幅药师佛说法图，这里选取的是主尊左侧的日光菩萨，与其相对的应为月光菩萨。日光菩萨是净琉璃世界无量无数菩萨众中的上首大菩萨，又称“日曜菩萨”“日光遍照菩萨”。在佛教造像中，常常以两位菩萨手中所执莲花之上的日轮和月轮来辨别身份，但在这幅壁画中，月光菩萨执莲，日光菩萨并无执物，且从服饰搭配上来看，除了偏袒左右肩的区别外，两位菩萨也无太大区别。

日光菩萨的形象明显受胡人基因的影响，体态丰腴，五官立体，眉目传神，两耳垂肩，侧面向佛，左手做出施予的姿态，右手放置于胸前，跣足站立在覆瓣莲花座上。菩萨戴仰月冠，束高髻，垂白色束发飘带，戴项饰、臂钏和手镯。服装的材料以丝织物为主，上身斜披披帛，束灰绿色围腰，下配短裙和长裙，随性而超然。披帛和长裙以土红色为底色，装饰着散点排列的石绿色十字花图案。装饰在身体相关部位的饰物，无不洋溢着初唐时期逐渐形成的淡雅而又富丽的艺术风格。画面的艺术处理上，线条圆润有力，色彩丽而不艳。

（文：刘元凤）

There is a picture of Bhaisajyaguru tableau on the south side of the east gate of the main chamber in cave 322 dating back to the early Tang Dynasty. Here is the sunlight Bodhisattva on the left side of the main Buddha, and the corresponding one is the moonlight Bodhisattva. Sunlight Bodhisattva is the prime Bodhisattva among countless Bodhisattvas in the world of pure glass world, also known as sun Bodhisattva or sunlight shining Bodhisattva. In Buddhist sculptures, the sun wheel and the moon wheel on the lotus held by the two Bodhisattvas are often used to identify their identities. However, in this mural painting, the moonlight Bodhisattva holds the lotus while the sunlight Bodhisattva does not hold anything. In terms of clothes, apart from the difference between the left and right bared shoulders, the two Bodhisattvas are not very different.

The image of sunlight Bodhisattva is obviously influenced by the gene of minority people. He has plump body, prominent facial contours, vivid eyebrows and eyes; two ear lobes reach to shoulders, facing to Buddha. The left hand means giving, standing on the lotus seat with bare feet and placing the right hand in front of the chest. This Bodhisattva wears moon crown, high bun, white hair bandeau, necklaces, armlets and bracelets. The material of the clothes is mainly silk fabric. The upper body is inclined covered by uttariya, wrapped a gray green waist wrap, a short antariya and a long antariya at the lower body, which is casual and supernatural. The uttariya and long antariya are decorated with scattered malachite green cross pattern with earth red background. The ornaments decorated on the relevant parts of the body are permeated with the elegant and rich artistic style gradually formed in the early Tang Dynasty. In the artistic treatment of the picture, the lines are mellow and powerful, and the colors are gorgeous but not flashy.

(Text by: Liu Yuanfeng)

图：刘元风　Picture by: Liu Yuanfeng

彩塑天王服饰

莫高窟初唐第322窟西壁龛北侧

The painted sculpture heavenly king's clothes

on the north side of the west niche in the main chamber of cave 322 dated to the early Tang Dynasty at Mogao Grottoes

唐时期的彩塑风格已趋写实，塑像的造型特征、精神风貌都生动传神、惟妙惟肖。虽然第322窟西壁龛内塑像还有隋代遗风，但造型与神态已显唐韵，尤其是龛北最外侧的天王塑像，雕塑手法细腻含蓄，人物形象栩栩如生，服装刻画真实自然，俨然是一位写实的唐代着甲武士形象。

唐初军队甲胄基本延续南北朝与隋朝的装备形制，这身天王塑像就是初唐甲胄的真实反映：自上而下分别穿戴兜鍪、颈甲、肩甲、护臂、胸甲、腿裙、胫甲、皮靴。有学者推断塑像上肩甲、腿裙上的彩条图案表现的是漆彩皮甲，皮甲是历史悠久的军装装备，商周时期制甲的主要材料就是皮革。唐时期的皮甲形制与明光甲相近，只是材质有异。

天王是佛教中的护法神，也是敦煌石窟艺术中频繁出现的形象。由于天王的职责是护持佛法、严防魔怪侵入，所以造型或怒目大吼、或深锁眉头，皆勇猛威严、坚毅果敢。第322窟西壁龛内的天王像则风格独特——身材高挑、身形纤瘦、脸型圆润、手指纤长，整体造型呈现出柔和优美的风格。从五官结构判断，龛内天王塑像有西域胡人的特征，这身天王像也因胡人的形象与温和的风格而在敦煌诸多刚猛的天王形象中独树一帜。

（文：李迎军）

The style of the painted sculptures in the Tang Dynasty had become realistic. The modeling features and spirit of the sculptures are vivid and lifelike. Although the sculptures in the west niche of cave 322 still have the Sui Dynasty style, their modeling and expression have shown the Tang Dynasty characteristics, especially the heavenly king in the north side of the niche. The sculpture modeling technique is delicate and subtle; the figure is lifelike, and the clothes is real and natural. It seems to be a real warrior of the Tang Dynasty.

The military armor in the early Tang Dynasty basically continued the equipment form of the Southern and Northern Dynasties and the Sui Dynasty. This painted sculpture heavenly king reflected the real armor in the early Tang Dynasty: from top to bottom, he wears the helmet, neck armor, shoulder armor, arm armor, chest armor, leg dhoti, shin armor and leather boots. Some scholars infer that the colored bar pattern on the upper shoulder armor and leg antariya of the sculpture is the feature of the painted leather armor, which was a military equipment with a long history. The main material of armor making in the Shang and Zhou Dynasties was leather. The shape of leather armor in Tang Dynasty is similar to Mingguang armor（明光甲）, but the material is different.

The heavenly king is the Dharma protector in Buddhism, which is also a common image in Dunhuang Grottoes. As the duty of the heavenly king is to protect the Dharma and prevent the invasion of demons and monsters, his image is always brave, serious and resolute. The heavenly king sculpture in the west niche of cave 322 has a unique appearance: tall, thin, round face and long fingers. The overall shape presents a soft and beautiful style. Judging from the facial features, this heavenly king has the characteristics of the Hu people in the Western Regions. This sculpture is unique among many other fierce heavenly kings' image in Dunhuang because of the gentle and Hu people style.

（Text by: Li Yingjun）

图：李迎军 Picture by: Li Yingjun

Buddha kasaya's trimming

on the north wall of the main chamber in cave 322 dated to the early Tang Dynasty at Mogao Grottoes

佛陀袈裟缘边图案

莫高窟初唐第322窟主室北壁

主室北壁佛陀袈裟缘边图案两条，因年代久远边饰表层颜料脱落、变色严重，底色均已变色为深褐色，表层纹样遗留青绿残迹。

敦煌壁画及彩塑着色所用的颜料，主要是天然矿物色颜料，以植物颜料和早期人工合成颜料为辅。敦煌研究院李最雄研究员在《敦煌莫高窟唐代绘画颜料分析研究》中对敦煌壁画中的颜料做了比较系统的科学分析，为我们今天绘制唐代服饰图案的色彩选择提供了科学有效的参照依据。隋唐五代的常见色为五色：红色、蓝色、绿色、棕黑色和白色，其中红色主要为朱砂与铅丹，蓝色主要是石青和青金石，绿色主要是石绿，棕色主要是棕褐色的二氧化铅，白色主要是碳酸钙矿物——方解石。

袈裟缘边图案由于装饰部位及装饰空间的限制，结构较为简单，纹样形象概括、均衡舒朗，色彩以石青、石绿色调为主，采用退晕的表现手法，色调浑厚富丽，体现了初唐边饰图案的装饰特征。

（文：高雪）

There are two patterns on the trimming of the Buddha's kasaya on the north wall of the main chamber. Because of the erosion of time, the pigments on the surface of the trimming decoration had fallen off and the colors changed seriously. The background color has changed to dark brown, and the surface pattern left traces of cyan and green.

The pigments used in Dunhuang murals and painted sculptures are mainly natural mineral pigments, supplemented by plant pigments and early synthetic pigments. Li zuixiong, a researcher of Dunhuang Research Academy, made a systematic and scientific analysis of the pigments in Dunhuang murals in his *Analysis and Research on the pigments of the Tang Dynasty paintings in Dunhuang Mogao Grottoes* (《敦煌莫高窟唐代绘画颜料分析研究》), which provides a scientific and effective reference for us to choose the colors of the Tang Dynasty clothes pattern today. The common colors in the Sui, Tang and Five Dynasties are red, blue, green, brownish black and white. Among them, red was mainly cinnabar and minium; blue was mainly azurite and lapis lazuli, green was mainly malachite green, brown was mainly brown lead dioxide, and white was mainly calcium carbonate mineral-calcite.

Due to the limitation of position and space, the structure of kasaya trimming pattern is relatively simple, and the pattern image is summarized, balanced and comfortable. The colors are mainly azurite and malachite green; color-gradation technique is also adopted. The colors are rich and solid, which reflects the decorative characteristics of the trimming pattern in the early Tang Dynasty.

(Text by: Gao Xue)

图：高雪 Picture by: Gao Xue

The trimming pattern of the painted sculpture Buddha's kasaya

on the west niche of the main chamber in cave 322 dated to the early Tang Dynasty at Mogao Grottoes

彩塑佛陀袈裟缘边图案

莫高窟初唐第322窟主室西壁龛内

由于袈裟袖缘图案残损，进而采用如下两个角度进行纹样复原推理研究。角度一：横向对比隋代及初唐各窟中相似的装饰图案，如第334窟、第372窟，推导此纹样的母题、骨架和装饰细节。角度二：研究第322窟整体设计风格，参照同窟的龛楣与藻井图案进行推理。综合研究后得到：第一种复原图案为初唐延续隋风的波状莲花纹：花冠为复合五瓣莲花，花萼为涡卷萼，纹样为连续波状“S”型骨式，主茎在转折处产生蘖枝，蘖枝上生出涡卷萼和莲花花苞。第二种复原图案为初唐新风的波状石榴莲花纹，花冠为复合面五瓣石榴花，花萼为石榴花萼，骨式为连续波状“S”型，主茎在转折处产生蘖枝，蘖枝上生出石榴花萼和莲花花苞。图案皆为经前代凝练后的程式化处理方式，构图具有婉转延展的动态美。图案以褐色为底色，以石绿填色、白色勾边，使得图案在光照有限的洞窟内更加突出了浮凸立体化的装饰效果。

（文：张博）

Because the kasaya sleeves trimming pattern is damaged, the following two angles are used for pattern restoration speculation. Angle 1: horizontally compare the similar decorative patterns in the caves of the Sui Dynasty and the early Tang Dynasty, such as cave 334 and cave 372, deduce the motif, frame and decorative details of this pattern. Angle 2: study the overall design style of cave 322 especially the niche lintel and caisson pattern of the same cave to speculate. The first restoration pattern is the wavy lotus pattern in the early Tang Dynasty, which continues the Sui style: the corolla is a compound five petaled lotus, the calyx is a scroll calyx, and the frame is a repeated wavy “s” shaped pattern. The main stem produces tillers at the turning point, and the scroll calyx and lotus bud grow on the tillers. The second restoration pattern is the wavy pomegranate lotus pattern in the early Tang Dynasty. The corolla is five petaled pomegranate flower with compound head, the calyx is pomegranate calyx, and the frame pattern is repeated wavy “s” shape. The main stem produces tillers at the turning point, and the pomegranate calyx and lotus bud grow on the tillers. The patterns are stylized after being condensed by the previous generation, and the composition has the twisted extension dynamic beauty. With brown background, malachite green filling and white trimming, the pattern highlights the three-dimensional decorative effect in the caves with limited light.

(Text by: Zhang Bo)

图：张博 Picture by: Zhang Bo

敦煌莫高窟初唐
第328窟

Dunhuang Mogao Grottoes Cave 328 of the Early Tang Dynasty

第328窟是敦煌莫高窟初唐时期的代表洞窟之一，为覆斗顶中小型窟。主室西壁开斜顶敞口龛，龛内彩塑一佛、二弟子、二菩萨、二供养菩萨，龛外另塑二胡跪供养菩萨（现存一身），龛外南侧的一身供养菩萨于1924年被美国华尔纳盗走，现藏于美国哈佛大学博物馆。龛顶绘倚坐佛说法图一铺，龛壁浮塑背光，两侧绘八弟子和二菩萨。前室北壁壁画为五代绘制，前室顶和西壁、甬道、主室窟顶和四壁的壁画均为西夏时期重绘。

此窟最精彩的部分是西壁龛内外的彩塑作品，基本保存了初唐原貌，是难得一见的唐代彩塑珍品。此窟彩塑均为成熟的圆塑，共包含八身人物，是彩塑艺术发展至极盛时期的经典组合。虽然人物众多，面貌各异，但作者运用塑、绘结合的手法，依据每一身塑像的身份、性格进行精准刻画。整铺塑像的造型比例准确，与真人等大，人物面相略长，颇具初唐时期的形象韵味。作者注重以身体姿态和面部神态反映不同类型人物的内心活动和内涵气质，将佛陀的庄严、阿难尊者的聪慧、迦叶尊者的苦修、菩萨的通达、供养菩萨的虔诚充分表现出来。

主室西壁龛顶的说法图描绘了彩云围绕众多菩萨、弟子，环绕倾听位于中央的弥勒佛说法的场景，龛内壁所绘平面的弟子像和菩萨像配合立体的彩塑穿插安置，相互呼应，以丰富的表现手段强化视觉重心、烘托礼拜观览时的氛围感，体现了作者在洞窟造像的整体规划和细节把握方面具有高超的能力与水平。

西龛内塑像的服装保存十分完整，服饰纹样清晰华丽、色彩明艳饱满，偏重大红、石绿和金色搭配，尤其是以阿难服装为代表的各类边饰，会采用大量金色绘制，展示了唐代服饰的华丽之风以及初唐时期高超的染织技艺面貌。裙身部分的服饰纹样为小团花，层层边饰中有不同的半团花。菩萨半裸上身，佩戴各式璎珞、首饰，并且愈加复杂，从中甚至可见类似盛唐的气象。塑像身后壁面和龛顶的人物服饰多有印度遗风，并且十分注重描绘大面积服装上的几何纹及小花纹饰。

（文：崔岩）

Cave 328 is one of the representative caves of Dunhuang Mogao Grottoes in the early Tang Dynasty. It is a small or medium-sized cave which has truncated pyramidal ceiling. The west wall of the main chamber has an open niche with an inclined top. Inside the niche, there are painted sculptures of one Buddha, two disciples, two Bodhisattvas and two offering Bodhisattvas. Outside the niche, there were two offering Bodhisattvas with Hu keel posture (only one left) , the one on the south side of the niche was stolen by Warner in 1924 and now collected in the Museum of Harvard University. The top of the niche is painted by a picture of sitting Buddha, the west wall of the niche is covered by back light relief, and the sides are painted with eight disciples and two Bodhisattvas. The murals on the north wall of the front chamber were painted in the Five Dynasties, while the murals on the top of the front chamber and the west wall, the corridor, the top of the main chamber and the four walls were repainted in the Western Xia period.

The most wonderful part of this cave is the painted sculptures inside and outside the west niche, which basically preserve the original appearance of the early Tang Dynasty, a rare treasure of the Tang Dynasty. The painted sculptures in this cave are all mature independent sculptures, including eight figures, the classic combination of the painted sculptures art in its heyday. Although there are many figures with different faces, the craftsman used the technique of combining sculpture and painting to accurately depict the identity and character of each sculpture. The proportion of the whole sculptures are accurate, as big as real people. The faces of the figures are slightly long and have the characteristic of the early Tang Dynasty. The craftsman focused on the body posture and facial expression to reflect the psychological activities and connotative temperaments of different types people, and fully shows the solemnity of Buddha, the wisdom of Ananda, the asceticism of Kasyapa, the peacefulness of Bodhisattvas, and the piety of offering Bodhisattvas.

On the top of the west niche of the main chamber, the painting depicts the scene of colorful clouds surrounding many Bodhisattvas and disciples, listening to the preaching of Maitreya Buddha in the center. The planar images of disciples and Bodhisattvas painted on the inner wall of the niche are interspersed with three-dimensional colored sculptures, echoing each other, strengthening the visual focus with rich means of expression, setting off the atmosphere of worship and viewing, reflecting the craftsman's excellent ability and high level overall planning for the cave sculptures and other details.

The clothes of the sculptures in the west niche are well preserved, with clear and gorgeous patterns, bright and colorful colors, and emphasis on the collocation of scarlet, malachite green and gold. In particular, all kinds of trimmings presented by Ananda's clothes are painted with a large amount of gold, which shows the luxurious style of the clothes in the Tang Dynasty and the superb dyeing and weaving skills in the early Tang Dynasty. The dress pattern of the dhoti part is small flower medallion, and there are different half flower medallion in the many layers of trimming decoration. Bodhisattvas are half naked, wearing many kinds of keyūra and jewelry and more complicated; from which we can even see the style is similar to the high Tang Dynasty. Most of the clothes on the wall behind the sculpture and on the top of the niche have Indian style, and paid great attention to depict the geometric patterns and small flower patterns on large areas of clothes.

(Text by: Cui Yan)

The pattern on the Buddha's kasaya and dhoti trimming

on the west niche of the main chamber in cave 328 dated to the early Tang Dynasty at Mogao Grottoes

彩塑佛陀袈裟缘边、裙边图案

莫高窟初唐第328窟主室西壁龛内

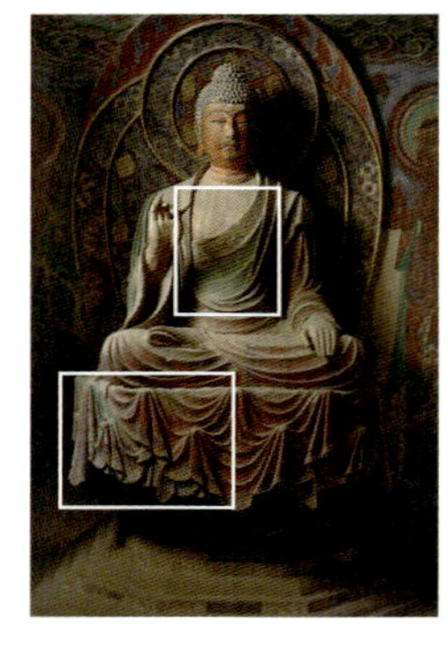

位于第328窟主室西壁龛内正中的主尊为彩塑结跏趺坐释迦牟尼佛，整座塑像按照严格的规范进行创作，除了面部和手的肤色经后代重绘，其余皆保持原作面貌。佛陀上身所着为“三衣”，即僧伽梨（Saṅghāṭi）、郁多罗僧（Uttarāsaṅga）、安陀会（Antarvāsaka），统称为袈裟。悬垂而下的袈裟下摆覆盖着座上的花瓣，形成了生动而有规律的波浪造型。袈裟一角露出厥修罗（Kusūla），即下裙的部分边缘。

这里选取整理绘制的是佛陀内层袈裟和裙子的缘边图案。袈裟主体为粉绿色，边缘为贴金忍冬纹。图案为条带式二方连续结构，枝条藤蔓呈波状卷曲，肥硕的四瓣叶上下交错而出，叶蒂处以三瓣叶填充，叶片造型体现出由纤细的北朝忍冬纹向丰硕的唐代卷草纹过渡的特点。图案配色简练，仅用金和石绿这两种色彩作为花色和底色，充分利用色彩结合造型所形成的正负形，营造出匀称而饱满的空间关系。裙子缘边为二方连续结构，主体为米字结构的半破式八瓣宝相花，在土红色的底色上以石绿色退晕的方式表现出花瓣的层次感，间以石青，以白色线条勾勒。彩塑佛陀服饰整体稳重庄严，在大面积的土红色袈裟中搭配石绿色的内衣和裙子，边缘饰以两条精致的图案，加上塑绘结合的手法表现出的衣褶变化和贴金装饰，质朴中不失华丽。

（文：崔岩）

Located in the center of the west niche of the main chamber in cave 328, the main painted sculpture is the Sakyamuni Buddha in lotus position. The whole sculpture was created according to strict standards; except the skin color of the face and hands had been repainted by later generations, the rest of the sculpture keeps the original appearance. The “three clothes” worn on the Buddha's upper body are called kasaya, namely, Saṅghāṭi（僧伽梨）, Uttarāsaṅga（郁多罗僧）, Antarvāsaka（安陀会）. The hanging lower hem of the kasaya covers the petals of the seat, forming a vivid and regular wave shape. A part of the Kusūla（厥修罗）can be seen, which is part of the trimming of the lower dhoti.

This is the trimming pattern of Buddha's inner Sankaksika and dhoti. The main body of kasaya is pink green, and the trimming is gilded honeysuckle pattern, the pattern is a stripe-shaped two in a group repeated structure; the branches and vines are wavy and curly. The full four petaled leaves crisscross up and down, and the pedicel is filled with three petaled leaves. The leaf shape reflects the transition from the slender honeysuckle pattern in the Northern Dynasty to the full scrolling grass pattern in the Tang Dynasty. The pattern is concise in color matching, which used only gold and malachite green as decors and background colors. The positive and negative shapes formed by color combination modeling are fully used to create a symmetrical and full spatial relationships. The trimming of the dhoti is a two in a group repeated structure, and the main body is a half eight petaled Baoxiang flower（半破式八瓣宝相花）with star shaped structure. On the earth red background, the petals show a sense of layers in the way of color-gradation technique by malachite green, and outlined by azurite and white lines. The overall clothes of the Buddha is solemn and steady, which used scarlet as the main color to match with malachite green Sankaksika and dhoti, and the trimming is decorated by two exquisite patterns, combined with sculpture and painting two kind of art languages, showing the pleats changes, along with gold decoration, making the clothes looks simple but magnificent.

（Text by: Cui Yan）

图：崔岩 Picture by: Cui Yan

The pattern on painted sculpture Ananda's shirt

on the west niche of the main chamber in cave 328 dated to the early Tang Dynasty at Mogao Grottoes

彩塑阿难偏衫图案

莫高窟初唐第328窟主室西壁龛内

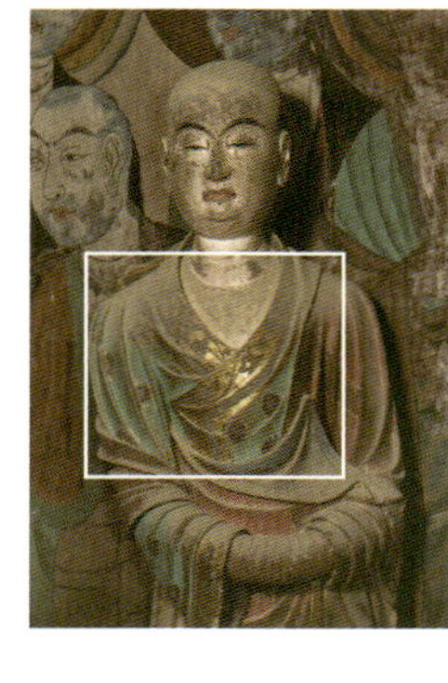

阿难尊者以“多闻第一”而著称，曾跟随佛陀四十五年听法布教。彩塑以丰腴的面庞和清秀的眉目刻画出他的聪慧专注，微微内倾的身姿表现了他正在侧耳倾听佛陀说法和精心思考的瞬间微妙。彩塑阿难尊者外披土红色袈裟，内着偏衫，下着裙，服饰色彩以土红色和石绿色为主，对比鲜明。

这里选取整理绘制的是彩塑的偏衫图案。偏衫图案为石绿底色上间错排列的六瓣散花纹，散花的花瓣共分三层，以石青色和土红色相互对比和呼应，花瓣均采用退晕手法表现，具有光芒的渐变感，在底色的衬托下显得熠熠生辉。偏衫领缘以贴金为地，上绘石青色和石绿色的卷草纹，纹饰主结构为波状曲线，交错而出的花叶一正一反，造型上带有石榴的特点，空间处理较为疏朗，体现出初唐时期卷草纹的特点。排列规律的散花纹和自由生长的卷草纹相得益彰，将彩塑人物衬托得更加华贵与生动。

（文：崔岩）

The venerable Ananda, known as “the most hearer”（多闻第一）, followed Buddha for 45 years to listen the Dharma. The painted sculpture used the plump face and delicate eyebrows to depict his intelligence and focused attention, and the slightly introverted posture shows the subtle moment when he is listening to the Buddha and thinking deeply. The sculpture is covered by earth red kasaya, Sankaksika and dhoti. The colors of the clothes are mainly earth red and malachite green with sharp contrast.

Here, we choose the pattern on the Sankaksika; it is a six petaled pattern with staggered arrangement on the malachite green background. The petals of the dispersive flower are divided into three layers, which are contrasted and echoed by the azurite and the earth red. The petals are all expressed by the technique of color-gradation, which has a sense of gradual change of light, also appears to be shining against the background color. The main structure of the pattern is wavy curves. The interlaced flowers and leaves have the characteristics of pomegranate; the space treatment is relatively sparse, which reflects the characteristics of the scrolling grass pattern in the early Tang Dynasty. The regular dispersive flower pattern and the free-growing scrolling grass pattern complement each other, making the figure more colorful and vivid.

(Text by: Cui Yan)

图：崔岩 Picture by: Cui Yan

彩塑迦叶尊者裙缘图案

莫高窟初唐第328窟主室西壁龛内

The pattern on painted sculpture Kasyapa's dhoti trimming

on the west niche of the main chamber in cave 328 dated to the early Tang Dynasty at Mogao Grottoes

彩塑释迦牟尼佛的左侧为迦叶尊者，他在佛陀十大弟子中号称“头陀第一”，是众弟子的首领。人物双眉紧蹙，眉骨、颧骨、胸骨突出，双手合十，是一位饱经风霜且富有德行的老者形象。彩塑迦叶尊者着袈裟、内衣和裙，袈裟以通肩式披着方式包裹全身，似乎在严寒之地修习苦行。

此处选取彩塑迦叶尊者的裙子边缘进行整理绘制。裙饰为二方连续结构，图案主体为半圆形宝相花纹，位于下列，半破式变体十字花为辅，位于上部。如果将此结构扩展为四方连续，便是唐代丝织品、金银器及建筑装饰中常见的团窠纹，根据主题的不同可分为宝相团窠、联珠团窠、动物团窠等。在敦煌莫高窟壁画中，这类图案除了用于服装边饰，还常常绘于藻井边饰和背光。此图案以红色为底色，主体花瓣为具有深、中、浅渐变层次的石绿色，间以石青色花瓣和红色花心，以白色线条勾勒外轮廓，排列为层叠的放射状，极具装饰性，为人物深沉的服饰增添了一抹鲜明的色彩。

（文：崔岩）

On the left side of the painted sculpture Sakyamuni Buddha is the venerable kasyapa, who is known as “the foremost in ascetic virtues”（头陀第一）among the ten disciples of the Buddha and was the leader of all the disciples. The figure's eyebrows are tight, the brow ridge, cheekbones and sternum are prominent, performing prayer mudra, who looks like an experienced and virtuous old man. The colored sculpture of venerable Kasyapa wears kasaya, Sankaksika and dhoti; the kasaya is wrapped around the whole body in the way of both shoulders covered, which seems that he is practicing asceticism in the cold area.

Here, we select the dhoti's trimming of the colored sculpture of the venerable Kasyapa to analysis. The dhoti decoration is a two in a group repeated structure. The main part of the pattern is a half Baoxiang pattern, which is located in the lower part; it supplemented by a variant semi cross pattern, which is located in the upper part. If we extended this pattern to four in a group repeated structure, then it became to the common Tuanke（团窠）pattern in Tang Dynasty's silk, gold and silver ware and architectural decoration. According to the different themes, it can be divided into Baoxiang Tuanke, Beads Tuanke, animal Tuanke, etc. In the murals of Mogao Grottoes at Dunhuang, this kind of pattern is not only used for clothes decoration, but also often painted on the trimming decoration of caisson and backlight. This design takes red as the background color; the main petals are malachite green with deep, medium and light gradients, and among them are azurite petals and red flower centers. The outline is white, arranged in a layered radial shape, which is highly decorative, adding a light color to the dark clothes of the figure.

（Text by: Cui Yan）

图：崔岩 Picture by: Cui Yan

The pattern on painted sculpture Manjusri Bodhisattva's antariya

on the north side of the west niche in the main chamber of cave 328 dated to the early Tang Dynasty at Mogao Grottoes

文殊菩萨裙身图案

莫高窟初唐第328窟主室西壁龛内北侧

文殊菩萨和普贤菩萨的服饰是第328窟五尊塑像中最为灿烂华丽的。敦煌石窟自十六国至元代的菩萨服饰虽然各不相同，但基本上为头顶束髻、戴宝冠、披天衣和披帛、饰璎珞、腰裹长裙，重点突出菩萨柔和优雅的形象。这两尊菩萨彩塑呈半跏坐像，头束高髻戴宝冠，面相略长，体态修长端庄，仪容高贵。文殊菩萨披披帛，普贤菩萨袒裸上身，均下着长裙垂于莲台，随着动态形成优美的衣纹，上绘精美的图案。

菩萨的裙饰图案主要包括土红色裙料上的四瓣及五瓣散花纹、两层锦裙边缘的一整二破式宝相花纹及二破式半宝相花纹、裙中菱格纹及膝盖处的圆形宝相花。将这些复杂的图案综合体现在同一锦裙的不同部位，体现了彩绘艺术家们的智慧。其中菩萨锦裙上交错排列的四瓣和五瓣散花纹与阿难尊者偏衫上的散花纹类似，花型小巧精致，排列疏密得当，花色和底色形成鲜明对比，显得生动活泼。菩萨锦裙边缘图案与前文提到的塑像袈裟或裙缘图案结构类似，但是花形结构和轮廓更加丰富，在米字结构的基础上衍生出十字形、六边形等，加以层层退晕的方法体现花瓣的递进关系，显得更加雍容华贵，在存世唐代宝相花纹织锦中也常见这样的色彩配置。

（文：崔岩）

The clothes of Manjusri Bodhisattva and Samantabhadra Bodhisattva are the most splendid among the five sculptures in cave 328. Although the clothes of Bodhisattvas in Dunhuang Grottoes from the Sixteen Kingdoms to the Yuan Dynasty are different, they all basically have a bun on the top of the head, wearing treasured crown, heavenly clothes and uttariya, keyūra and long antariya around the waist; all are trying to highlight the graceful and elegant image of Bodhisattvas. The two painted sculptures Bodhisattvas are heroine posture（半跏坐）seated figures with a high bun and crown on their head. Their faces are slightly long; bodies are slender and dignified, and their appearances are noble. The Manjusri Bodhisattva is covered by uttariya, and the Samantabhadra Bodhisattva bare his upper body. They all wear long antariyas and droop down on the lotus seat, forming the dynamic feeling of clothes wrinkles, decorated with exquisite patterns.

The patterns of Bodhisattva's antariya decoration mainly include four petals and five petals dispersive flowers（散花纹）scattered on the earth red area, one whole and two halves and two halves of Baoxiang flower pattern on the brocade antariya two layers trimmings, rhombic pattern on the antariya and Baoxiang flower pattern on the knee. The arrangement of these complex patterns in different parts on one brocade antariya reflects the wisdom of the craftsman. The staggering arranged four petals and five petals dispersive flower patterns on the Bodhisattva's brocade antariya are similar to those on Ananda's Sankaksika; they are small and delicate, and the arrangement is appropriate. The pattern and background are in sharp contrast, which makes it lively. The trimming pattern of the Bodhisattva's brocade antariya is similar to the painted sculpture kasaya or antariya trimming pattern mentioned above, but the flower structure and outline are more complex. The star structure developed into the cross and hexagon shape, and the blossom of petals reflected by color-gradation technique, which is more elegant. This kind of color configuration is also common in the existing Tang Dynasty Baoxiang pattern brocade.

（Text by: Cui Yan）

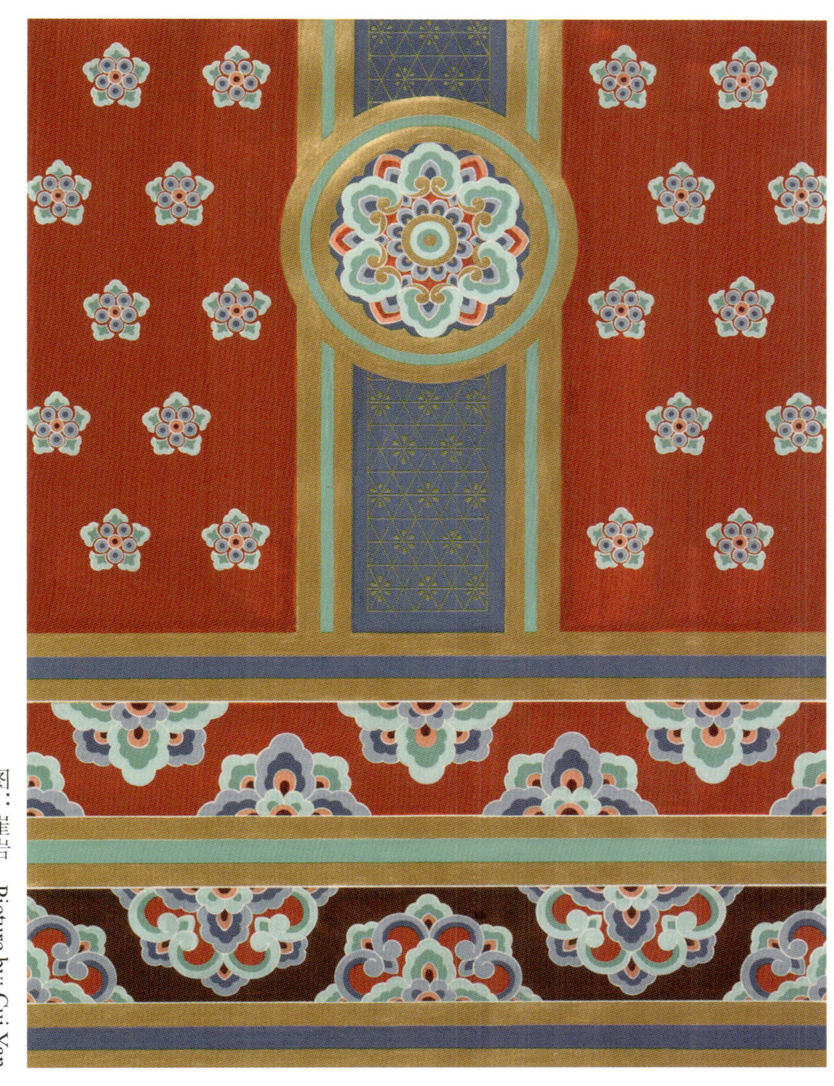

图：崔岩 Picture by: Cui Yan

普贤菩萨裙身图案

莫高窟初唐第328窟主室西壁龛内南侧

The pattern on the painted sculpture Samantabhadra Bodhisattva's antariya

on the south side of the west niche in the main chamber of cave 328 dated to the early Tang Dynasty at Mogao Grottoes

两尊菩萨服饰图案的特别之处在于裙中菱格纹及膝盖处圆形宝相花的搭配，这样的装饰不同于其他几尊塑像。精细的石青地金线菱格纹贯通全裙，与边饰相交，在菩萨的膝盖部位以开光的方法绘出圆形的石绿色宝相花纹，六出花瓣融合了如意云头的造型，与边饰的二方连续图案既有联系又有区别，相互呼应，巧妙而新颖。因这两处装饰与周围底料的色彩和装饰风格存在差异，中间有色条分隔，特别是膝盖处图案与菩萨的坐姿和身形配合得恰到好处，如同定位设计一样。这在敦煌壁画和彩塑菩萨的服饰图案中并不常见，相信与织造水平的提高和彩塑艺术家的创造密切相关。

两身菩萨的裙饰图案整体华丽繁复，结构独特，在大面积的土红色底上点缀色阶分明的石绿、石青，并用贴金和白线勾勒外轮廓来调适对比，达到造型和色彩的和谐统一。

（文：崔岩）

The special feature of the two Bodhisattvas' dress patterns is the rhombic pattern on the antariya matching with the Baoxiang flowers at the knees, which is different from other sculptures. The exquisite malachite green and gold line rhombic pattern runs through the whole antariya and intersects with the trimming ornament. The round malachite green Baoxiang pattern is drawn on the Bodhisattva's knee with the method of opening light; the six petals are integrated with the shape of Ruyi cloud head; they are connected with and different from the two in a group repeated patterns of the trimming ornament, echoing with each other, ingenious and novel. Because of the differences in color and decorative style between the two decorations and the surrounding base materials, there is a colored stripe in the middle to separate, especially the pattern at the knees is just right with the sitting posture and body shape of the Bodhisattva, just like position design（定位设计）. This is not common in Dunhuang murals and dress patterns of painted Bodhisattva, which is believed to be closely related to the improvement of weaving technique and the creativity of the craftsman.

The overall design of the two bodies Bodhisattva's antariya is gorgeous and complex with unique structure. It used a large area of earth red as background, and decorated with clear malachite green and azurite, and the outer contour is outlined by gold and white lines to adjust the contrast, so as to achieve the harmonious unity of shape and color.

（Text by: Cui Yan）

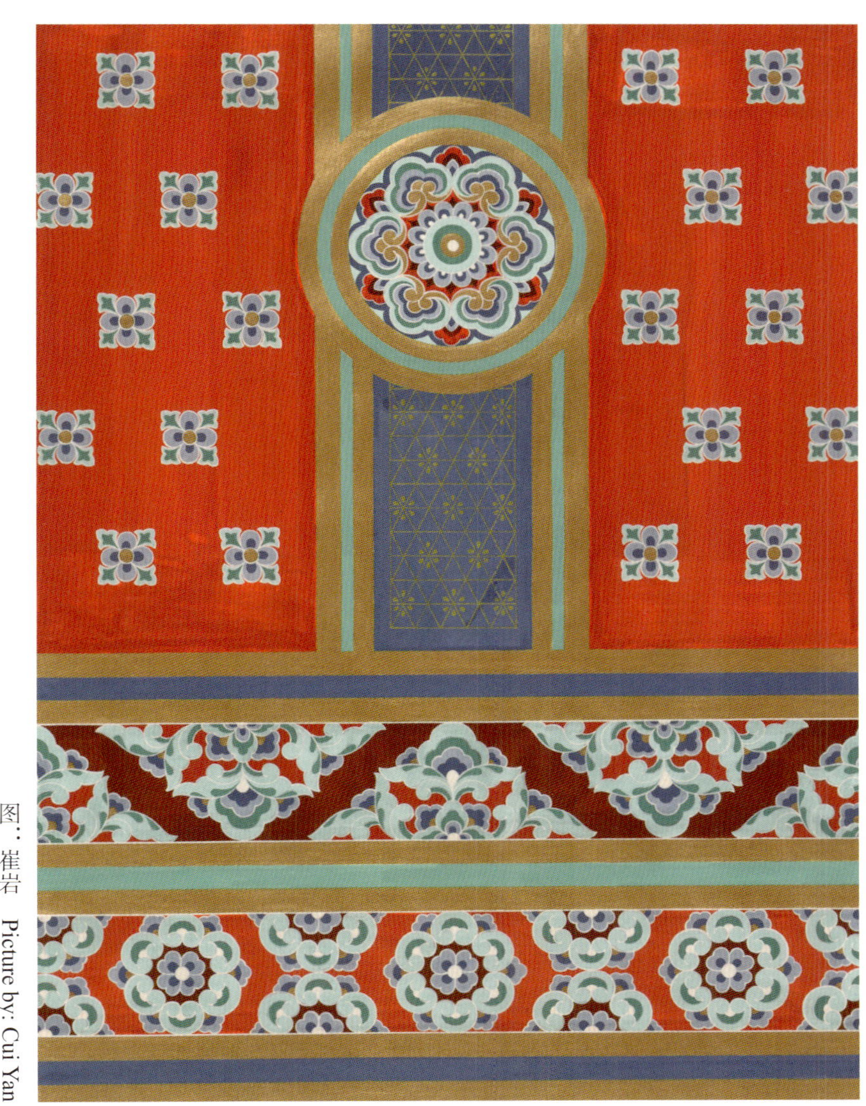

图：崔岩 Picture by: Cui Yan

The pattern on painted offering Bodhisattva's short antariya, long antariya

on the south side of the west niche in the main chamber of cave 328 dated to the early Tang Dynasty at Mogao Grottoes

供养菩萨短裙、长裙图案

莫高窟初唐第328窟主室西壁龛内南侧

壁画上菩萨裹石绿色围腰，下着短裙和长裙，装饰于短裙下摆的珠玉缀饰衬托出人物的动感。图案主体为典型的菱形网格纹，菱格内点缀十字白色联珠，赭石底色衬托出了青、绿色带的鲜明脉络。初唐沿袭隋代的细线镂金描画的方法，像刺绣的线镶嵌一般，附着在纹样上，起进一步刻画形象的作用。敦煌壁画中菩萨的形象特征之一即身披璎珞，珠玉缀饰对配饰的细致刻画表现，体现了随佛教一同传来的印度装饰习俗。

菱格这一图案骨架早在商代就已出现于织物的几何纹上，根据历史文献和实物表明，战国丝织工艺取得了很大的成就，几何纹多数是以菱形为变体的主体形。构成方法主要有两种，一种是直接用变体菱形纹作均匀排列，另一种是用连续的几何纹网作骨架，再在其中填入与之相适应的几何形或变体几何纹，点缀其间的菱形适合纹样随菱形内空间大小而设计填充，各种几何元素大小、聚散、疏密均有变化。

菩萨的短裙图案细致精美、造型巧妙、色彩和谐，作为古代配饰的一种，其深厚的民族文化性为今天的设计同样提供了创意灵感来源。

菩萨的长裙图案为红色底上分布小散花，散花分大小两种，上下交错，朴素美观，有活泼的层次感和韵律感，壁画中仅见朵花的白色轮廓线。这种图案形式早在北朝就已出现，实物如新疆吐鲁番阿斯塔纳墓出土的西凉时期蓝地蜡缬绢，其花朵同样是分大小两种，小朵花排成菱格，大朵花填于格心，交叉排列。日本藏唐代蜀红锦中心的小花也是相同组织形式。汉唐蜀锦名闻天下，丝绸古道上出土了很多汉唐遗物均产自蜀地，由此可以窥得这种几何散花的样式在早期织物中已有呈现，并一直延续至唐代。红色地体现了唐代图案的用色特点，将朱色衬托或点缀在青绿色纹样间，常以赭、红调和，形成与青绿纹样的补色关系，并用金色与黑白等中性色衬托出图案的金碧辉煌。

（文：高雪）

In the mural, this Bodhisattva is wrapped in malachite green waist wrap, wearing short antariyas and long antariyas on the lower body. The jade beads decoration at the bottom of the short antariya complement the dynamic character of the figure. The main body of the pattern is a typical rhombic pattern, which is dotted with cross white beads, and the ochre background color complement the distinct stripes of azurite and green. The early Tang Dynasty followed the Sui Dynasty's method of fine line engraving drawing（细线镂金描画）, like the inlay of embroidery line attached to the pattern, playing a role in further depicting the image. One of the characteristics of Bodhisattva in Dunhuang murals is the meticulous depiction of the accessories like keyūra and jade beads, following the Indian dressing custom.

As early as in the Shang Dynasty, the rhombic pattern appeared on the fabrics as a kind of geometric patterns. According to historical documents and historical relics, great achievements have been made in silk weaving technology during the Warring States period. Most of the geometric patterns are rhombic based variations. There are two main ways to form the rhombic pattern: one is to arrange the rhombic pattern evenly, the other is to use the repeated geometric pattern net as the frame, and then fill in corresponding geometric pattern or variant geometric pattern. The fillings in the rhombic pattern will be adjusted according to the size of the space, including the size, aggregation, dispersion and density.

Bodhisattva's antariya pattern is exquisite in design, ingenious in shape and harmonious in color. As a kind of ancient accessories, its profound national culture significance also provides a source of creativity inspiration for today's design.

The pattern on Bodhisattva's red long antariya is scattered dispersive flowers. There are two kinds of scattered dispersive flowers: one is large and the another is small. They are staggered up and down, simple and beautiful, with a lively sense of layers and rhythm. Only the white outline of the dotted flower can be seen in the mural, this kind of pattern has appeared as early as in the Northern Dynasties. For example, the blue background wax dyed silk fabrics（蓝地蜡缬绢）dating back to the Xiliang period unearthed from Astana tomb in Turpan, Xinjiang, has two kinds of flowers: the small flowers are arranged in a rhombic lattice, and the large flowers are filled in the lattice and arranged in cross way. Shu (Sichuan province) brocade of the Han and Tang Dynasty was famous all over the world. Many silk relics of the Han and Tang Dynasty were unearthed on the ancient Silk Road, all of which were produced in Shu. From this, we can see that this geometric pattern of scattered dispersive flowers had appeared in the early fabrics and had lasted to the Tang Dynasty. Red background color reflects the color characteristic of the Tang Dynasty pattern. Ochre and red were often blended to form a complementary relationship with blue-green patterns, and neutral colors such as gold and black-and-white were used to set off the resplendence of the patterns.

（Text by: Gao Xue）

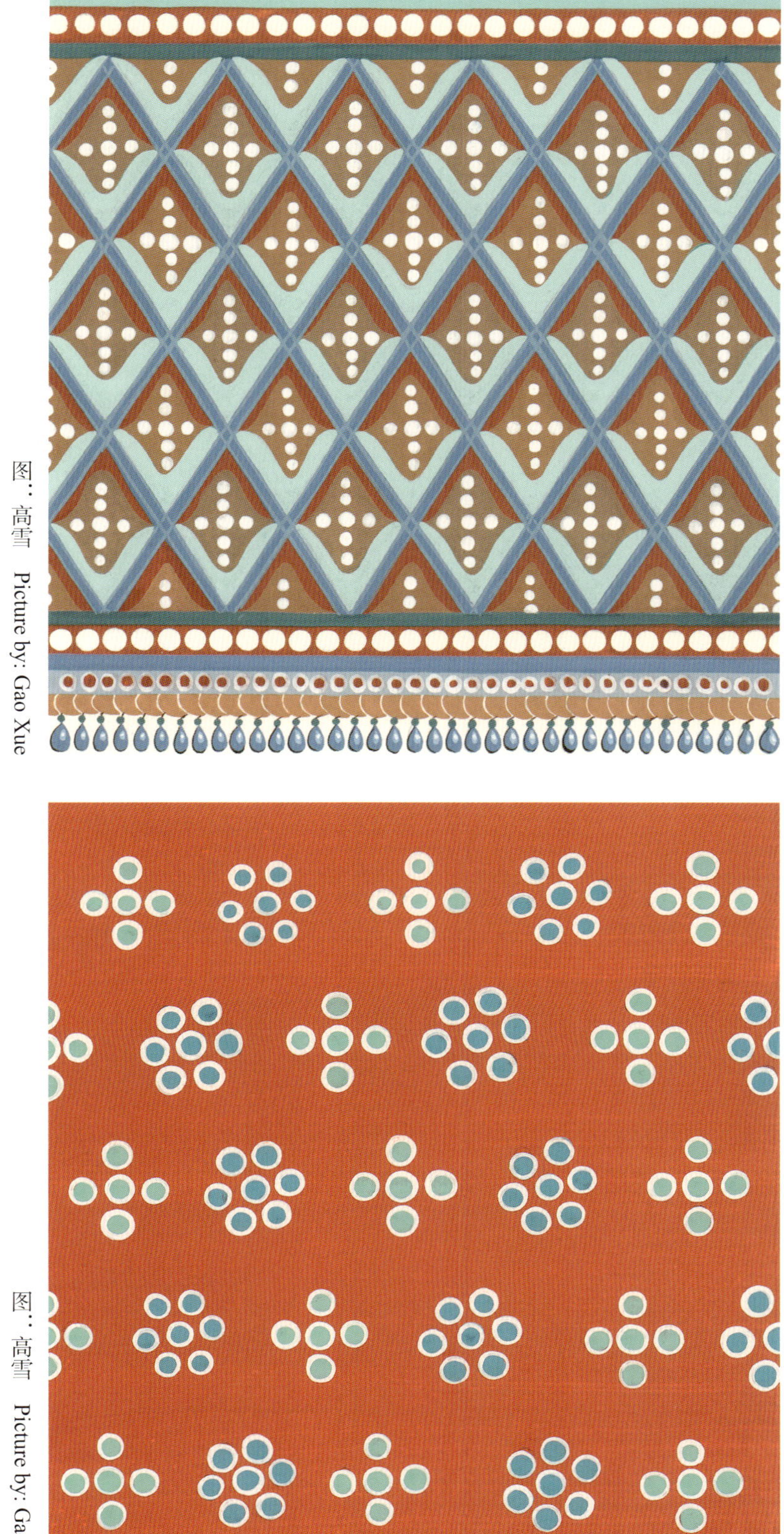

图：高雪 Picture by: Gao Xue

图：高雪 Picture by: Gao Xue

The pattern on painted offering Bodhisattva's uttariya

on the south side of the west niche in the main chamber of cave 328 dated to the early Tang Dynasty at Mogao Grottoes

供养菩萨披帛图案

莫高窟初唐第328窟主室西壁龛内南侧

菩萨上身斜披披帛，图案整体分布呈二方连续，小菱格内饰有十字联珠，并组成内框。以红色为底，外框边缘有蓝红褐色相间的色带，装饰白色联珠，底色加描白线，颇具唐代特色。菱格具有整齐一致、平衡对称的形式美感，在绞缬、蜡缬、夹缬等古代染织工艺中较易实现。重彩加描白线是初唐装饰图案常用的手法，虽然朱红、石青、石绿均已变色，仍体现出当时锦绣衣饰的精致和绚丽。

菱格纹作为一种显性的几何纹样，被大量应用在中国古代提花织物中。古代河南信阳长台关楚墓出土的战国菱纹绢，因为纹样外形似战国时代盛行的漆羽觞（耳杯），因此文献上称这类菱形纹为“杯纹”，是中国菱形纹样的早期艺术表现形式，见于《释名・释彩帛》：“有杯文，形似杯也”。在各地出土的战国、秦汉的罗、绮类织物上多数为这类纹样，其中湖南长沙马王堆西汉墓出土的“杯纹”形象最为典型，与战国的菱形纹样一脉相承。根植于中国传统装饰纹样的菱格纹，经过历史的检验流传至今，仍常见于建筑、织物和各类工艺美术作品的装饰当中。

（文：高雪）

This Bodhisattva's upper body is obliquely covered by uttariya, the overall pattern is two in a group repeated. Inside the little rhombuses have cross dots, with red as the background, the trimming of the outer frame has blue, red, brown color bands which are decorated by white dots, and the background has white lines, which is quite the Tang Dynasty style. It was easy to have this pattern in ancient dyeing and weaving processes such as tie-dyeing, wax-dyeing and clamp-dyeing. Solid color combined with white lines was a common method in decorative pattern in the early Tang Dynasty. Although vermilion, malachite green and azurite colors have changed, this still reflect the delicacy and brilliance of clothes at that time.

Rhombic pattern was very popular and widely used in ancient Chinese jacquard fabrics（提花织物）. The rhombic pattern silk unearthed from the Chu tomb in Changtaiguan, Xinyang, Henan Province dating back to the Warring States is called "cup pattern" in the literature because its shape is similar to the lacquer cup（ear cup）prevailing in the Warring States period. It was an early artistic form of rhombic pattern in China. The record can be found in *Shi Ming-Shi Cai Bo*《释名・释彩帛》（the explanation of the colored silk names）: "there is a kind of cup pattern, shaped like a cup". Most of the Luo（罗, silk gauze）and Qi（绮, damask）fabrics unearthed from the Warring States period, Qin and Han Dynasties have this kind of pattern, among which the "cup pattern" unearthed from Mawangdui（马王堆）tomb in Changsha, Hunan Province is the most typical, which is the same linage of the Warring States Period. Rooted in traditional Chinese decorative patterns, the rhombic pattern, which has been handed down through historical examination, is still common in the decoration of buildings, fabrics and various arts and crafts nowadays.

（Text by: Gao Xue）

图：高雪 Picture by: Gao Xue

敦煌莫高窟初唐

第329窟

Dunhuang Mogao Grottoes Cave 329 of the Early Tang Dynasty

第329窟是敦煌莫高窟初唐时期的代表洞窟之一，为覆斗形窟。主室藻井为莲花飞天井心，四披各绘伎乐天三身，下绘千佛。西壁开斜顶敞口龛，龛内彩塑一佛、二弟子、四菩萨，除佛像的身躯、衣饰及金刚座为初唐原作外，其余皆为清代重修。龛顶绘夜半逾城（南侧）、乘象入胎（北侧）、伎乐天、雷神等，龛壁正中绘化佛火焰背光，两侧各绘弟子四身及人非人、飞天、化生等，下北侧绘婆薮仙，南侧绘鹿头梵志。龛外南北两侧上部绘千佛，下部绘莲花童子各二身。龛下绘供案、供器和供养菩萨六身，多已模糊。主室南壁绘阿弥陀经变，下部五代绘供养比丘尼三身和女供养人二十三身。北壁上部绘弥勒经变一铺，下部为五代绘男供养人十七身。东壁门上绘说法图四铺，南北两侧各绘说法图一铺、供养牛车马群和男女供养人画像。前室和甬道壁画多为五代时重修绘制。

南北两壁通壁的巨幅经变画题材反映了初唐时期盛行的净土思想，极尽想象力地描绘了西方净土世界和弥勒净土世界的美妙庄严。图中佛、菩萨、天人穿插绘制于宫殿楼阁之中，周围栏楯水榭、桥梁连通，七宝铺地，宝幢高耸。天空有飞天散花和不鼓自鸣的仙乐，地下有明净通透的莲花宝池。整体构图较隋代复杂丰富，形象描绘更加细腻，在清新淡雅中不失绮丽之风，体现出初唐逐渐向盛唐风格转变的趋势。

西龛顶“乘象入胎”与“夜半逾城”中相关人物形象分别着印度式和中原式服装，西壁龛外西侧童子服饰十分生动。北壁弥勒经变中群组佛菩萨服饰与东北侧服饰都比较注重小花朵的装饰，对于世俗纺织品面料纹样的模仿十分细致，如不同的染色、印花等。北壁下部和东壁下部的几组世俗人物服装反映了当时的流行风尚，可以从中看到比较原貌的初唐服饰搭配，整体色彩以青绿和大红搭配为主。如同初唐同时期多个洞窟一般，东壁菩萨服装多表现裤装。南壁色彩氧化严重，人物的服装服饰仅能分辨出基本的造型，具体的细节和装饰纹样已经模糊不清。

（文：崔岩）

Cave 329 is one of the representative caves with truncated pyramidal ceiling in Dunhuang Mogao Grottoes dating back to the early Tang Dynasty. The caisson in the main chamber has lotus and flying Apsaras in the center, each of the four slopes has three performers and Thousand Buddha motif. The west wall has an open niche with an inclined top; there are painted sculptures of one Buddha, two disciples and four Bodhisattvas in the niche. Except the body, clothes and Vajra seat of the Buddha are original works of the early Tang Dynasty, the rest part were remade in the Qing Dynasty. The top of the niche is painted the picture of the Great Departure (south side), the Great Conception (north side), performers, thunder God, etc. The center of the niche wall is painted with the flame backlight of the Buddha; on each side are painted four bodies of disciples and non-human, flying Apsaras, birth out of lotus, etc. On the lower north side is painted the Vasistha（婆薮仙）, and on the south side is painted the Mrgasirsa（鹿头梵志）. On the north and south outside of the niche, and the upper part are painted Thousand Buddha motif, and the lower parts of the both sides are painted two bodies of lotus children respectively. Below the niche are painted the offering table, the offering utensils and the six bodies of Bodhisattva but all blurred. The south wall of the main chamber is painted Amitabha Sutra tableau, and the lower part is painted three bodies of nuns and twenty-three bodies of female donors. The upper part of the north wall is painted Maitreya Sutra tableau, and the lower part is painted male donors in the Five Dynasties. There are four pictures of preaching scenes above the east gate, both south wall and north wall have a picture of preaching scene together with male and female donors, horses and cattle on the lower part. Most of the murals in the front room and corridor were repainted in the Five Dynasties.

The huge sutra tableau on the north and south walls reflect the prevailing pure land thought in the early Tang Dynasty; they manifest the wonderful and solemn world of the Western Pure Land and Maitreya pure land with great imagination. Buddha, Bodhisattva, heaven and celestial beings are interspersed in the palaces and pavilions, surrounded by balustrades, waterside pavilions and bridges, seven treasures on the floor and towering buildings. In the sky, there are flying Apsaras scattering flowers and self playing musical instruments, and clear lotus pond in the ground. Compared with the Sui Dynasty, the overall composition is more complex and rich, and the image description is more delicate, we can see prosperity in simple and elegance, reflecting the trend of the early Tang Dynasty gradually changing to the high Tang Dynasty style.

The related characters in “the Great Conception” and “the Great Departure” on the top of the west niche are dressed in Indian style and Central Plains style respectively; the children’s clothes on the outside of the west niche are very vivid. In the Maitreya Sutra tableau on the north wall, the Buddhas and Bodhisattvas’ clothes and the clothes on northeast side, the ancient craftsman payed more attention to the small flowers decoration. The imitation of secular textile fabric patterns are very detailed, such as different dyeing, printing effect and so on. Several groups of secular people clothes on the lower part of the north wall and the east wall reflect the prevailing fashion at that time. From which we can see the original clothes collocation of the early Tang Dynasty and the overall color is mainly green and red. Like many caves at the same period in the early Tang Dynasty, Bodhisattva’s clothes on east wall mostly depicted the pants. The color oxidation of the south wall is serious, only the basic shape of the characters’ clothes can be distinguished, and the specific details and decorative patterns have blurred.

(Text by: Cui Yan)

供养菩萨服饰

莫高窟初唐第329窟主室南壁

The offering Bodhisattva’s clothes

on the south wall of the main chamber of cave 329 dated to the early Tang Dynasty at Mogao Grottoes

这是初唐第329窟主室南壁阿弥陀经变中右上侧的一身供养菩萨，虽然原壁画中菩萨的肌肤已经变色，但是通过动态能够感受到他凝重而专注的神情。画面表现了菩萨侧面立于覆瓣莲花座上，双手合十，视线集中在手中的莲苞上，正缓步行进的瞬间姿态。

菩萨戴宝珠冠，梳高髻，束发飘带从背后垂落，戴璎珞、臂钏、手镯和脚镯。镶嵌着宝珠的璎珞在身后垂成环形，兜住长裙，形成变化丰富的衣褶。其上系结的长带正背两面分别为石绿色和土红色，自肩后绕至双臂后垂下，随着菩萨的走动飘落在身后。敦煌唐代壁画中的服饰常将对比色用于织物的正反面，这样的配置显得服饰色彩更加鲜明艳丽。菩萨束腰带，下着松身长裙，裙腰处翻折出石绿色的内里，形成双层叠穿的效果，结构表现较为灵活，具有实际穿着的真实感。菩萨的佩饰与衣服相得益彰，整体色调沉稳雅致，其中穿插点缀土红、石绿等明丽的色彩，并通过虚化的团花纹花砖作为人物的背景衬托，共同营造出行云风动、飘飘欲仙的视觉美感。

（文：刘元凤）

This is a Bodhisattva on the upper right side of the Amitabha Sutra tableau south wall of the main chamber in cave 329 dating back to the early Tang Dynasty. Although the skin color of the Bodhisattva in the original wall painting has changed, his dignified and attentive expression still can be felt through the posture. This picture shows the Bodhisattva profile standing on the petal covered lotus seat. His hands folded, eyes focused on the lotus bud, walking slowly.

This Bodhisattva wears jewel crown, high bun, hair streamers falling from the back, keyūra, armlets, bracelets and anklets. The keyūra inlaid pearls hang as a circle behind the body and hold the long antariya, forming a variety of pleats. The two sides of the long belt are malachite green and earth red respectively; it goes from shoulder back to arms and falls behind the body. In Dunhuang murals of the Tang Dynasty, the contrast color was often used in the front and back of the fabric, which made the clothes color more bright and gorgeous. The Bodhisattva wears a belt and a long loose antariya, the waist wrap turns over and shows malachite green inside lining, forming a double-layer overlapping effect; the structure is flexible and has a realistic feeling of actual wearing. Bodhisattva’s ornaments and clothes complement each other; the overall tone is calm and elegant, interspersed with bright colors such as earth red and malachite green, and the virtual round pattern tiles are used as the background to create the visual beauty of movement and nobility.

(Text by: Liu Yuanfeng)

图：刘元风 Picture by: Liu Yuanfeng

The female donor's clothes

on the south side of the east wall in the main chamber of cave 329 dated to the early Tang Dynasty at Mogao Grottoes

女供养人服饰

莫高窟初唐第329窟主室东壁南侧

初唐第329窟东壁南侧说法图中这身女供养人的服饰可以说是当时贵族女性中最具时尚感的代表。她蛾眉凤眼，墨发椎髻，面部神态温婉而虔敬，体态丰满而柔美，手持长梗含苞初放的莲花，跪于方毯之上。类似构图和动态的供养人像在敦煌藏经洞出土的《树下说法图》（八世纪初）中亦能看到，可见这是当时流行的供养形式。

这身女供养人位于整幅壁画中主尊右侧胁侍菩萨莲花座的下方，身形较小，与佛、菩萨高耸的形象对比悬殊，与之相对的还有一身穿黄色圆领袍、手持柄香炉的男供养人，可见这是以家族供养的形式而创作的壁画。她的服饰造型简约得体，上着紧身圆领窄袖小衫，套半臂，衣着紧身贴体，显露出丰腴的体态。下着高束腰间色长裙，蓬松而舒适。这种高束腰的间色长裙具有明显的修身效果，是唐代女子的流行穿着。供养人身上没有过多的饰品，却显得自然、娴静，落落大方。在画面的艺术表现上，用线刚劲奔放，疏密虚实得当；色彩配置和谐有序，层次分明。

（文：刘元风）

On the south side of the east wall of cave 329 dating back to the early Tang Dynasty, the dress of the female donor in the picture can be said to be the most fashionable representative noble woman at that time. She has moth eyebrows and phoenix eyes, black hair and a bun. Her face is gentle and reverent, and the body is plump and soft. She kneels on the carpet, holding a lotus flower. Similar composition and dynamic images of the donors can also be seen in the painting of *Preaching Under Trees* （《树下说法图》, early 8th century） unearthed from Dunhuang Library Cave. It can be seen that this was a popular form of offering at that time.

This female donor is located on the right side of the main image in the whole mural, below the lotus seat of the attendant Bodhisattva. Her body is small, which is quite different from the towering image of Buddha and Bodhisattva. In opposite, there is a male donor wearing a yellow round collar robe and holding a censer. It can be seen that this mural is created in the form of family offering. Her dress style is simple and appropriate, with a tight round neck narrow long sleeves shirt, half sleeves shirt outside, and the clothes fit the body well, revealing a plump body. The lower body wears high waist striped color antariya, fluffy and comfortable. This kind of high waist striped color long antariya has obvious slim effect and was popular for woman in the Tang Dynasty. There are not many jewelry on the donor, but they make this lady look natural, quiet and confident. In the artistic expression of the picture, the lines are vigorous and unrestrained, and the density is appropriate; the color configuration is harmonious and orderly, and the layers are clear.

（Text by: Liu Yuanfeng）

图：刘元风 Picture by: Liu Yuanfeng

The children's clothes on both outsides of the west niche in the main chamber of cave 329 dated to the early Tang Dynasty at Mogao Grottoes

童子服饰 莫高窟初唐第329窟主室西壁龛外两侧

第329窟主室西壁龛外南、北两侧上下各画化生童子两身，除身体皮肤部分氧化脱色严重外，服装相对完整，将儿童憨态可掬、顽皮好动的个性表现得出神入化。

在这四身化生童子中出现了两种唐代的典型儿童服饰"涎衣"与"裲裆"。涎衣也称"繄袼""围涎"，为古代儿童颈部常用的东西，一般为圆形或者方形，戴在颈部，后部开口可以系住，它的实际用处相当于今天的围帕。裲裆是与兜肚形制相似的一种服饰，其形制类似今天的吊带背心，前胸、后背各有一片衣襟，无袖，肩部用两条带子连接。唐代儿童在夏日尤喜着裲裆，在周昉的《戏婴图》中有相似的儿童形象，证明此窟壁画童子形象的写实风格。

西壁龛外南侧上方的化生童子穿着红色镶石绿色边涎衣，发型推测为三搭头。左手搭在莲花骨朵上，右手放在体侧，左脚抬起跃跃欲试向前迈出，右脚轻踏在一朵盛开的莲花中心。下方的化生童子穿着红色镶边裲裆，脚蹬软靴，左手托着莲花，右手拇指插在靴筒里。

（文：吴波）

Outside the west niche of cave 329, the south and north side are painted two incarnation children（化生童子）respectively. Apart from the serious oxidation and discoloration of the skin, the clothes are relatively complete, showing the children's charming and playful personality.

There are two typical kinds of children's clothes of the Tang Dynasty in this painting: Xianyi（涎衣）and Liangdang（裲裆）. Xianyi is bib commonly used on children's neck in ancient times, they are usually round or square in shape, worn on the neck and can be tied by an opening at the back. Their practical use is equivalent to today's bib. Liangdang is a kind of clothes similar to today's suspenders vest, with a piece of cloth at the front and back, no sleeves, and two suspenders at the shoulders. Children especially like Liangdang during summer in the Tang Dynasty. There are similar images of children in Zhou Fang's *Xi Ying Tu*（《戏婴图》, the picture of children playing）, which proves the reliability of this painting.

The incarnation boy on the south side of the west niche is dressed in red bib inlaid with malachite green trimming, and his hairstyle is supposed to be triple pieces; the left hand is on the lotus stem, the right hand is on the side of the body, the left foot is raised, eager to step forward, and the right foot lightly steps on the center of a blooming lotus. The incarnation boy below is wearing a red suspenders vest, soft boots, holding lotus in his left hand, and right hand thumb in his boot.

（Text by: Wu Bo）

图：吴波 Picture by: Wu Bo

童子服饰

莫高窟初唐第329窟主室西壁龛外两侧

The children's clothes

on both outsides of the west niche in the main chamber of cave 329 dated to the early Tang Dynasty at Mogao Grottoes

西壁龛壁北侧上方的化生童子穿着红色镶绿边裲裆，右臂向下绕过一株莲蓬，左手向下握着一支莲苞，左脚踏在莲蓬中心，右腿向上抬起，头部微微向左斜望。下方的化生童子梳垂髫，近代称这种发式为“一抓椒”“冲天炮”。童子身着红色涎衣，脚蹬橘色软靴，里子皆为石绿色。童子右手轻握莲花杆，左手轻触莲花花苞，右脚踩踏在莲蓬中心，左腿抬起轻踏于一支莲花杆上。整理绘制时，弱化了童子身边的莲花造型，着意还原突出童子的服饰、生动的体态和调皮的表情。

（文：吴波）

At the top of the north side of the west niche wall, the incarnation boy wears a red Liangdang with green trimming, his right arm down around a lotus seedpod, his left hand down holding a lotus bud, his left foot in the center of the lotus seedpod, his right leg up, and his head slightly tilted to the left. The boy below combed his hair in one braid, in modern times, this kind of hairstyle is called “pigtail”. This boy wears red bib and orange soft boots with malachite green lining, holding the lotus stem by his right hand, touches the lotus bud by his left hand, standing on the center of the lotus pod by his right foot, and raises his left leg to step on a lotus stem. When drawing the picture, the lotus shape around the boy is weakened, and the clothes, vivid posture and naughty expression of the boy are paid more attention.

(Text by: Wu Bo)

图：吴波 Picture by: Wu Bo

Flying Apsarases' clothes

on the top of the west niche in the main chamber of cave 329 dated to the early Tang Dynasty at Mogao Grottoes

飞天服饰

莫高窟初唐第329窟主室西壁龛顶

常书鸿先生在《敦煌飞天》一文中指出飞天“在印度，梵音叫她犍闼婆，又名香音神，是佛教图像中众神之一”，而佛经中多以“诸天”“天人”“天女”来称呼飞天。

按佛经所示，飞天的职能有三：一是礼拜供奉，表现形式为双手合十，或双手捧花果奉献；二为散花施香，形式为手托花盘、花瓶、花朵，或拈花散布；三为歌舞伎乐，表现形式为手持各种乐器，演奏、舞蹈。敦煌壁画中的飞天，按布局大致可分为藻井飞天、平棊飞天、人字披飞天、龛顶及龛外飞天、背光飞天、法会飞天、环窟飞天，按其造型有童子飞天、六臂飞天、裸体飞天，按其职能有伎乐飞天、散花飞天、供养飞天、托物飞天等。

莫高窟第329窟西壁龛顶为佛传故事乘象入胎的情景，飞天的服装形制与初唐的菩萨类似，上身袒露胸背，无斜披，下着及踝长裙，腰部向外翻出，显示异色的面料。左飞天双手捧莲花，右飞天双手弹箜篌，都赤裸双足，脖颈佩戴璎珞。飘带正反异色，于身长数倍的尺寸飘于空中，随意卷曲，与唐代民间女性所戴披帛十分相似。这种裸露上身的服饰样式更多地保留着印度服饰的较为原初的样貌。飞天体态优美，以曲线造型、线条流畅的飘带穿插其间，疏密有致，在层层叠叠的云纹的衬托下，与点状的漫天香花共同营造出一片祥瑞之气。

（文：张春佳）

In *Dunhuang flying Apsaras*（《敦煌飞天》）, wrote by Mr. Chang Shuhong, pointed out that flying Apsaras “In India, Sanskrit calls her Gandharva（犍闼婆）, also known as sweet sound God（香音神）, which is one of the gods in Buddhist images”. In Buddhist scriptures, flying Apsaras is often referred to as “celestials（诸天）”, “celestial beings（天人）” and “celestial woman（天女）”.

According to the Buddhist scripture, flying Apsaras has three functions: one is to worship, which is expressed in the form of holding hands together or holding flowers and fruits; the second is to scatter flowers, which is in the form of holding flower trays, vases, flowers or scattering flowers; the third is performers, which is expressed in the form of holding various musical instruments, playing and dancing. According to the layout, the flying Apsaras in Dunhuang murals can be roughly divided into Caisson flying Apsaras, Pingluo flying Apsaras, gabled ceiling flying Apsaras, niche top and outside flying Apsaras, backlight flying Apsaras, preaching flying Apsaras and around cave flying Apsaras. According to their shapes, there are children flying Apsaras, six arms flying Apsaras and naked flying Apsaras. According to their functions, there are music flying Apsaras, scattering flower flying Apsaras, offering flying Apsaras and holding object flying Apsaras.

At the top of the west niche of cave 329 of Mogao Grottoes is the scene of the Buddha's story of the Great Conception. The clothes design of flying Apsaras is similar to that of Bodhisattvas in the early Tang Dynasty. The upper body is naked, without uttariya, and the ankle length antariya is worn below, the waist part clothes is turned out to show the different color of the fabric. The left Apsaras holds lotus by both hands, and right Apsaras plays Konghou（箜篌, similar to harp）by both hands. Both of them are barefoot and wear keyūra around their neck. The sashes are different in color, longer several times than the length of the body, floating in the air, curled freely, which is very similar to the Tang Dynasty woman's uttariya. This kind of dress style with bare upper body is more close to the original appearance of Indian dress. Flying Apsaras' body is beautiful, decorated by curved sashes interspersed with each other. With the background cloud patterns, it creates an auspicious atmosphere together with the dotted fragrant flowers.

（Text by: Zhang Chunjia）

图：张春佳 Picture by: Zhang Chunjia

供养菩萨裙身图案

莫高窟初唐第329窟主室北壁

The pattern of offering Bodhisattva's antariya

on the north wall of the main chamber in cave 329 dated to the early Tang Dynasty at Mogao Grottoes

图案取自初唐第329窟主室北壁弥勒经变中位于主尊右侧的大妙相菩萨的裙身。大妙相菩萨在众菩萨的环绕下，结跏趺坐于莲花台上，面部微侧向佛。菩萨上身披石绿色披帛和透明的天衣，着红色长裙，裙子饰以朵花纹和绿色散点纹。主纹为朵花纹，花心为绿色椭圆环，外围以白色羽状花瓣环绕，呈放射状，造型上具有菊花的特点。花纹之间的空隙以绿色点状纹填充，与朵花纹交错呈四方连续排列。纹样灵动俏丽，花地分明，清地排列，错落有致。

（文：苏芮）

The design is taken from Bodhisattva antariya on the right side of the main image in the Maitreya Sutra tableau on the north wall of the main chamber in cave 329 dating back to the early Tang Dynasty. This Bodhisattva is surrounded by other Bodhisattvas, sitting in lotus position on the lotus platform with his face slightly facing to the Buddha. The Bodhisattva's upper body is covered with malachite green uttariya and transparent heavenly clothes, wearing a long red antariya decorated with flower patterns and scattered green dots patterns. The main pattern is a flower; the center has a green oval ring, and the periphery is surrounded by white radial pinnate petals; the shape has the characteristics of chrysanthemum. The gap between the patterns are filled by green dots pattern, which are crisscrossed with the flower pattern four in a group repeated arrangement. The pattern is beautiful and well arranged, and the contrast with the background is clear.

(Text by: Su Rui)

图：苏芮 Picture by: Su Rui

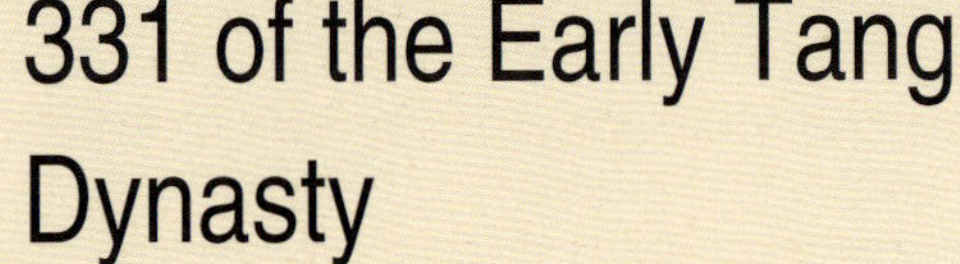

敦煌莫高窟初唐 第331窟

Dunhuang Mogao Grottoes Cave 331 of the Early Tang Dynasty

莫高窟第331窟为李家窟，李家是敦煌的大姓之一，世代为官。据第332窟中《李君莫高窟佛龛碑》记载的信息推测，第331窟为李达所修，时间约为七世纪中期。西龛内彩塑残存一些唐代服饰纹样的痕迹，尤其是龛外南北两侧天王服饰上依稀可见大团花装饰纹样。东壁门上法华经变中的人物服饰清晰细致，设色典雅。虽为整块色彩，但用流畅的线进行细节表现。南北两壁下部分别有成组女、男供养人画像。南北两壁阿弥陀经变中的人物服饰多以长线来表现披帛和服装结构。

（文：张春佳）

Cave 331 of Mogao Grottoes is the Li family's cave. Li family was one of the great surnames in Dunhuang and has been working in government for generations. According to the information recorded in the stele of *Li Jun Mogao Grottoes*（《李君莫高窟佛龛碑》）in cave 332, it is speculated that cave 331 was built by Li Da（李达）in the middle of the seventh century. There are some traces of Tang Dynasty clothes patterns on the painted sculptures in the west niche, especially the decorative pattern of big flower medallion pattern on the heavenly king's clothes on the north and south outside of the niche. Above the east gate, the clothes of the characters in the lotus sutra tableau are clear and meticulous, and the colors are elegant. Although they are solid color, they have smooth lines for details. In the lower part of the north and south walls, there are groups of female and male donors respectively. In the tableau of Amitabha Sutra on the north and south walls, people's clothes are mostly used long lines to show their uttariya and clothes structure.

（Text by: Zhang Chunjia）

男供养人服饰

莫高窟初唐第331窟主室北壁下

The male donor's clothes

on the lower part of the north wall in the main chamber of cave 331 dated to the early Tang Dynasty at Mogao Grottoes

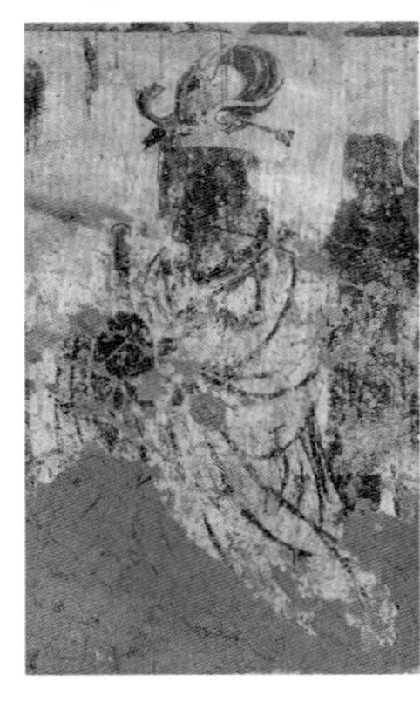

图中所绘为北壁下部男供养人，服冠，冠顶向后旋卷，前部有博山、梁。交领袍服，手持笏板，不露手。壁画中所绘冠形类似于孙机先生在《中国古舆服论丛》中所提及的新疆柏孜克里克壁画中绘制的通天冠。从宋代武宗元《朝元仙仗图》中南极、东华天帝君的冠看，与其有相似之处——冠顶、博山、导等部位。目前来看比较近似的均为帝王冠。《后汉书·明帝纪》李注引《汉官仪》："天子冠通天，诸侯王冠远游，三公、诸侯冠进贤三梁。"

李家源自西凉王李氏一族，李暠随母亲移居敦煌后，曾任敦煌太守。该段壁画损毁严重，此供养人约处于第四单元的位置。如果按照李家李暠以及第332窟窟主李克让任左玉钤卫谷府校尉的情况推及，《李君莫高窟佛龛碑》记载第331窟为李克让父亲李达所开，也记载了几位相关的祖辈。第五十八行为李达，其名字前面提到过四位：远祖頠、显祖昭、曾祖穆，还有一位空缺，如此，第331窟北壁下部如果从西向东四位依次排列的话，可能第四位就是这位空缺，时任"隋大黄府上大都督、车骑将军"的李家先祖。如果本部分壁画绘制的正是这位大都督，那么其职务为掌管一方的行政长官，故而以夸张的高级别冠的形态描述其形象倒也是情理之中。另外，服色或为褪掉——唐代典籍制度中对于服装颜色的规制与此不符，无论从一品至七品，均不为白色。

（文：张春佳）

The picture shows a male donor in the lower part of the north wall, wearing a crown, with the crown curling backward and Boshan（博山）and Liang（梁）in the front, holding the Huban（笏板）and keep hands inside sleeves. The shape of the crown in the mural is similar to that in the mural of Bozikrik in Xinjiang, which was mentioned by Mr. Sun Ji in *Chinese ancient clothes*（《中国古舆服论丛》）. There are also some similarities compare to the crowns of the Heavenly Emperors Nanji（南极）and Donghua（东华）in the painting of *Chaoyuanxianzhangtu*（《朝元仙仗图》, three heavenly emperors visiting the lord of heaven）by Wu Zongyuan of Song Dynasty, such as the Guanding（冠顶）, Boshan（博山）, Dao（导）parts and so on. As far as we can see, this crown is very close to ancient imperial crown. Li's annotation in the book of the *late Han Dynasty, the chronicle of the Ming emperor*（《后汉书·明帝纪》）, quoted *the officer's regulation of the Han Dynasty* as saying: "the emperor wears Tongtian（通天）crown, dukes or princes wear Yuanyou（远游）crown, the marquis wears Jinxiansanliang（进贤三梁）crown."

Li family originated from the king of Xiliang Li family. After Li Hao（李暠）moved to Dunhuang with his mother, he was once the prefecture chief of Dunhuang. The mural is seriously damaged, and the donor is located in the fourth unit. According to the Li family history, Li Hao and Li Kerang（李克让）, the owner of cave 332, cave 331 was sponsored by Li Kerang's father, Li Da（李达）, and several related ancestors are recorded in the monument of *Li Jun Mogao Grottoes stele*（《李君莫高窟佛龛碑》）. The name of Li Da, appeared in the 58th line, also refers four people before his name: the remote ancestor Wei（頠）, the near ancestor Zhao（昭）, great-grandfather Mu（穆）, and there is still a vacancy. If the four people in the lower part of the north wall in cave 331 are arranged from west to east, the fourth one maybe the vacancy, the ancestor of the Li family who was then the "great commander and general of the Sui Dahuang house". If this part of the mural depicts this governor, then his position was the chief executive in charge of one area, so it is reasonable to depict his image in the form of exaggerated way who wears an crown. In addition, the color of the clothes may faded, because the regulation of clothes color for officers in the Tang Dynasty has no match with this; no matter from rank one to rank seven, there was no white clothes.

（Text by: Zhang Chunjia）

图：张春佳　Picture by: Zhang Chunjia

敦煌莫高窟初唐

第332窟

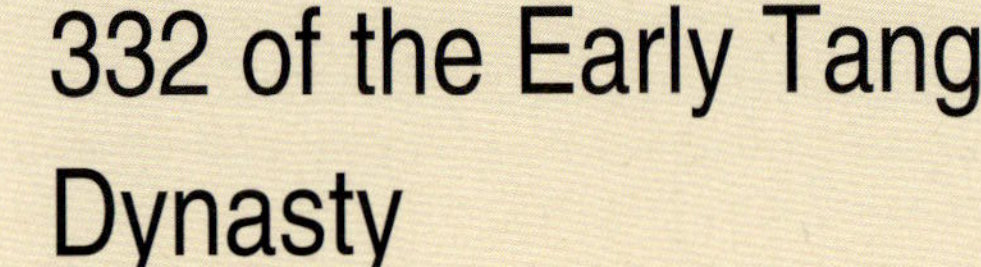

Dunhuang Mogao Grottoes Cave

332 of the Early Tang

Dynasty

第332窟因原存武周圣历元年（698年）所建《李君（克让）莫高窟佛龛碑》一方，故又名“李克让窟”“圣历窟”，是建于初唐时期的人字披中心塔柱窟。主室中心方柱东向面塑一佛二菩萨像，上绘文殊、普贤各一铺，南、西、北向面分别绘卢舍那佛、药师佛、灵鹫山说法图。窟顶前部两披和后部平顶均绘千佛，西披中部绘释迦多宝佛一铺。主室西壁开横长圆券形龛，龛内塑涅槃像一身，龛内外壁画均为涅槃变中的经典情节及供养菩萨等人物，与塑像主题相呼应。南壁和北壁前部各塑一佛二菩萨立像一铺，上绘赴会佛二铺，南壁后部绘涅槃经变，北壁后部绘维摩诘经变，下分别为女供养人和男供养人。主室东壁门上绘珞珈山观音一铺，门南绘一佛五十菩萨图，门北绘灵鹫山说法图一铺，下部分别绘三身供养比丘。门沿存五代绘菩萨二身，前室和甬道均存五代和元代壁画。

北壁维摩诘经变中有大量表现世俗人物服饰的场面，如帝王、大臣、各国王子等。南北两壁服饰注重以线表现结构，流畅丰富。对于服杂裾垂髾天女的形象来讲，以西魏时期的典型服饰特征为范本，这样的形制一直流传下来——当天女出现在唐前期各个时期的壁画中时，这样的服装形制均十分相像。此时，南北两壁服饰的小花朵纹样也与同时期其他洞窟相类似，展示着初唐的纺织品印染工艺。

此窟保留的《李君（克让）莫高窟佛龛碑》十分珍贵，内容述及莫高窟的创建年代及此窟的开凿情况等，对研究莫高窟的创建和发展具有关键意义。窟内壁画内容丰富，相互之间具有明显的呼应关系，如南壁的涅槃经变与西壁龛内彩塑涅槃像合璧便是一铺关于《大般涅槃经》的完整演绎，这在莫高窟壁画中是罕见的例子。虽然窟内大部分壁画已经变色，但所幸主室西壁基本上保持了初唐壁画绚丽鲜艳的原貌，尤其是龛下所绘形态各异的供养菩萨和珍禽异兽，线条遒劲有力，色彩鲜艳如初，可以作为考察研究同时期洞窟艺术风格的重要参考。

（文：崔岩）

Cave 332 is a gabled central pillar cave built in the early Tang Dynasty,because there was a stele in the cave named *Li Jun* (*Kerang*) *Mogao Grottoes Stele* [《李君（克让）莫高窟佛龛碑》], the time inscription was 698 A.D., the first year of the Shengli period of the Wu Zhou Dynasty, so this cave is also known as Li Kerang grotto or Shengli grotto. In the center of the main chamber, there are one Buddha, two painted sculptures of Bodhisattva on the east side of the square column, including one painted Manjusri Bodhisattva and one painted Samantabhadra Bodhisattva on the wall. On the south, west and north side, there are pictures of Vairocana Buddha, Bhaisajyaguru and Griddhakuta mountain preaching. The front two slopes and the back flat ceiling are painted with Thousand Buddha motif, and the middle part of the west slop is painted with a picture of Sakya and Prabhutaratna Buddha. In the west wall of the main chamber, there is a long horizontal oval shaped niche, in which there is a Nirvana sculpture. The murals inside and outside the niche are the classic plots of Nirvana tableau and offering Bodhisattva, which echo the theme of the sculpture. One Buddha and two Bodhisattvas painted sculptures are made at the south wall and north wall each, at the same time the upper parts of the sculptures are painted with two pictures of attending Buddhas. The back part of the south wall is painted with Nirvana Sutra tableau, and the back part of the north wall is painted with Vimalakirti Sutra tableau. The lower parts of the both walls are female and male donors respectively. On the east wall of the main chamber, there is a picture of Guanyin on Luojia mountain（珞珈山）which is painted above the east door, one Buddha and fifty Bodhisattvas are painted on the south side of the door, a picture of Buddha preaching on Griddhakuta hill is painted on the north side of the door, and both the lower parts have painted three offering bhikkhus. There are two Bodhisattvas painted in the Five Dynasties around the gate, and the front room and corridor have murals of the Five Dynasties and the Yuan Dynasty.

In the Vimalakirti Sutra tableau on the north wall has a lot of scenes showing the clothes of secular characters, such as emperors, ministers, princes and so on. The clothes patterns on the north and south walls emphasized in lines, fluency and richness. As for the image of celestial woman wearing Zajuchuishao（杂裾垂髾, a kind of clothes style, which has triangle hemline on the antariya, long belt draping behind body）, the typical clothes features of the Western Wei Dynasty have been inherited down. When celestial woman appeared in the murals of various periods in the early Tang Dynasty, the shapes of such clothes were very similar. At this time, the small flower patterns of the clothes on the north and south walls are similar to other caves of the same period, showing the textile printing and dyeing technique of the early Tang Dynasty.

The *Li Jun* (*Kerang*) *Mogao Grottoes stele* preserved in this cave is very precious. The content describes the founding time of the Mogao Grottoes and the excavation of this cave, which is the key information to study the creation and development of the Mogao Grottoes. The murals in the grottoes are rich in content and have obvious corresponding relationship with each other. For example, the combination of the nirvana Sutra tableau on the south wall and the painted Nirvana sculpture in the west niche are a complete interpretation of the *great Nirvana Sutra*, which is a rare example in the murals of the Mogao Grottoes. Although most of the murals' color in this cave have changed, fortunately, the west wall of the main chamber basically retains the original gorgeous appearance of the early Tang Dynasty. In particular, the murals painted below the niche have different forms of offering Bodhisattvas, rare birds and animals with strong lines and bright colors, which can be used as an important reference for the study of cave's artistic style of the same period.

(Text by: Cui Yan)

供养菩萨服饰

莫高窟初唐第332窟主室西壁龛下

The offering Bodhisattva's clothes

under the west niche of the main chamber in cave 332 dated to the early Tang Dynasty at Mogao Grottoes

初唐壁画常在四壁下部绘制供养菩萨像，第332窟也遵循了这样的绘制范式，在主室西壁涅槃龛下绘十四身供养菩萨像。供养菩萨姿态不尽相同，形象俊美，服饰清晰，线描与色彩都保存尚好。

这里选取其中的两身供养菩萨进行整理绘制。其中一身供养菩萨神态从容，双目微闭，俯视前方，手持柄香炉，轻步前行。头戴花冠，身上佩戴璎珞、臂钏和手镯。上身斜披石绿色披帛，翻折处露出红色的里面，色彩对比极为鲜明。胯间饰以彩色段状绦带和红色围腰，下穿镶以红色缘边的石青色短裙，以及透明质感的石绿色阔腿收口长裤，裤子上装饰着红色的十字花纹。

（文：刘元风）

In the early Tang Dynasty, the craftsman often painted images of Bodhisattva on the lower part of the four walls. Cave 332 also followed this tradition, it has 14 Bodhisattvas below the nirvana niche on the west wall of the main chamber. The postures of these offering Bodhisattvas are not the same; the images are beautiful, and the clothes are clear, the line drawing and colors are well preserved.

Here we choose two of them to draw and analyse. One is the Bodhisattva with a calm look, eyes slightly closed, overlooking the front, holding a censer and walking forward. He wears a wreath on his head, necklace, armlets and bracelets on his body. The upper body is obliquely covered by malachite green uttariya, and the inside is red exposed at the folds, with bright color contrast. The crotch is decorated with colorful belt and red waist wrap, the azurite antariya with red trimming covered the lower body, and the transparent malachite green wide legs necked trousers are decorated with red cross pattern.

(Text by: Liu Yuanfeng)

图：刘元风 Picture by: Liu Yuanfeng

供养菩萨服饰

莫高窟初唐第332窟主室西壁龛下

The offering Bodhisattva's clothes

under the west niche of the main chamber in cave 332 dated to the early Tang Dynasty at Mogao Grottoes

另一身供养菩萨榜题为“越三界菩萨”，她双眸低垂，神情宁静，左手托玻璃宝珠盘，右手持鲜花。菩萨头顶花冠，佩戴璎珞、臂钏和手镯，上身披披帛，正反两面为石绿色和红色，装饰着红色的四簇点纹，腰束彩色段状绦带，下着红色围腰、石绿色短裙和褐色的阔腿收口长裤，裤子上点缀着四瓣花纹。菩萨的首饰和飘动的披帛相辅而行，整体上给人以静中有动的美感。

两身供养菩萨肤色晕染自然生动，所穿服饰是初唐时期供养菩萨中较为常见的搭配和造型，色彩以红色、石绿、石青、赭石为主。原壁画经历了烟熏与磨损，已经较为斑驳，尤其头冠部分比较模糊，因此参考同时期的供养菩萨宝冠进行推断和绘制，但是菩萨的整体服饰仍然保留着鲜艳的本色，十分难得。

（文：刘元风）

The other one is entitled “越三界菩萨”(the Bodhisattva crossing three realms). His eyes are low and the expression is quiet. He holds a glass jewel tray in his left hand and flowers in his right hand. This Bodhisattva wears a wreath on his head, a necklace, armlets and bracelets. His upper body is covered by uttariya, the front and back sides are malachite green and red, decorated with four clusters of red dots. He wears colorful ribbons around his waist, a red waist wrap, a malachite green antariya and a brown wide legs necked trousers. The trousers are decorated with four petals flower patterns. Bodhisattva's jewelry and flying uttariya complement each other, giving people a sense of beauty of moving in stillness.

The skin color of the two Bodhisattvas are natural and vivid, and the clothes they wear are the common collocation and modeling of Bodhisattva in the early Tang Dynasty. The colors are mainly red, malachite green, azurite and ochre. The original murals have experienced smoke and touch, and have mottled, especially the head crown part is fuzzy. Therefore, we used other offering Bodhisattva crown in the same period to infer and draw, but the whole dress of the Bodhisattva still retains its bright original color, which is very rare.

(Text by: Liu Yuanfeng)

图：刘元风 Picture by: Liu Yuanfeng

The prince's clothes of different countries

on the north wall of the main chamber in cave 332 dated to the early Tang Dynasty at Mogao Grottoes

各国王子听法图服饰

莫高窟初唐第332窟主室北壁

各国王子听法图是维摩诘经变图的重要部分，壁画中出现的各国王子形象不仅是当时中国社会与中亚、西亚、南亚，乃至欧洲各国密切交往的实证，也为研究当时各国服饰提供了难得的图像资料。由于缺乏其他的文字与图像佐证，西域的民族迁徙频繁而复杂，年代久远，而且壁画也日渐斑驳模糊，所以目前无法清晰辨别每一个人物的国别与民族属性。

第332窟主室北壁维摩诘经变中这位正在听法的王子形象，头戴尖顶织锦帽，着织锦缘饰窄袖长袍，足蹬尖头乌皮靴。北方草原民族习惯戴尖顶帽，唐时期对这种尖顶卷沿的帽子统称为“胡帽”，胡服、胡帽在唐朝曾经广为流行，并对唐代服装造型发展产生了重要影响。

现在这部分壁画尽管略显斑驳，但服装服饰的造型特点与配色关系尚可分辨，只是长袍的织锦饰边图案已识别不清，绘图整理时，参照初唐洞窟中的图案绘制了长袍边饰。

位于听法王子队列前排的王子头缠长布带，着窄袖绿色长袍，领、襟、衣摆、袖口都装饰有红色织锦饰边，足蹬尖头乌皮靴，支靴的钩状尖头极具特色。由于壁画已经斑驳，目前无法分辨这位王子的发饰，参照其他洞窟绘制的各国王子礼佛图中的王子形象，绘制整理为光头造型。

（文：李迎军）

The pictures of princes listening to Dharma are an important part of Vimalakirti's Sutra tableau. The images of princes in the murals are not only the evidence of the close contact between Chinese society and Central Asia, West Asia, South Asia, and even European countries at that time, but also provide rare image information for studying the clothes of various countries at that time. Due to the lack of literature and image proof, and the ethnic migration in the Western Regions was frequent and complex, also the time is old, and the murals have mottled and blurred, so it is impossible to clearly distinguish the nationality and ethnics attribute to characters at present.

In the Vimalakirti Sutra tableau on the north wall of the main chamber of cave 332, the prince who is listening to the Dharma is wearing a pointed tapestry hat, a narrow sleeves long robe and pointed black leather boots. The northern grassland people used to wear pointed hat. In the Tang Dynasty, this kind of rolled brim pointed hat was called "Hu hat" (Minority's hat). Hu clothes and Hu hat were popular in the Tang Dynasty, and had an important impact on the development of clothes modeling in the Tang Dynasty.

Although this part of the mural is slightly mottled, the relationship between the modeling characteristics and color matching of clothes are still distinguishable, but the pattern of the tapestry trimming of the robe is not clear. When drawing the robe trimming, we used the general patterns of the early Tang Dynasty caves.

The prince in the front of the group wears a long cloth band on his head and green robe with narrow sleeves. His collar, hem and cuffs are decorated with red tapestry. His feet are wearing pointed black leather boots, and the hook like tips of the boots are very distinctive. As the murals have mottled, it is impossible to distinguish the prince's hair ornaments at present. Referring to the prince's image in other caves, the hairstyle is arranged into a bald head.

(Text by: Li Yingjun)

图：李迎军 Picture by: Li Yingjun

各国王子听法图服饰

莫高窟初唐第332窟主室北壁

The prince's clothes of different countries

on the north wall of the main chamber in cave 332 dated to the early Tang Dynasty at Mogao Grottoes

各国王子听法图中的这身形象头戴莲花状小冠，上身内穿方心曲领中单，外套阔袖袍，腰间以阔丝带系结，带端垂于前，长至膝下，裙前垂蔽膝，脚穿高齿履。

由于氧化变色与脱落，现在壁画上的人物形象与色彩已经较难完整辨认，绘制整理时，根据相关线描资料与文献的记载，主要复原了曲领中单与皮肤的颜色，并画清了头冠的形状。头上的小冠因外形仿若盛开的莲花，称“莲花冠”，莲花冠在唐代时已在世间流行，在后世仍沿袭其制久为流传。“中单”是套在宽大的袍服内穿着的内衣，通常以白色纱罗或布帛制作。唐时期的方心曲领是附在中单上的一种装饰领，即在中单上衬起半圆形的硬衬，可以使领部凸起。唐代方心曲领是七品以上官员礼服上装饰的领型，通常与袍服、大袖襦、中单搭配使用，多为白色。宋代曾经“遵循唐制”而在朝服交领上戴方心曲领，但宋代的这种上圆下方、形似璎珞的饰件，与唐代方心曲领的造型已是天壤之别。

（文：李迎军）

This prince wears a lotus shaped crown on his head, a square heart curved collar Zhongdan（方心曲领中单）inside, a broad sleeves robe as coat, and a wide silk belt around the waist. The belt ends are hanging down to the front of his knees. The Bixi（a piece of rectangular cloth hanging in front of legs）covers the legs over the knees, and his feet wearing high tooth shoes.

Because of oxidation and discoloration, it is difficult to completely identify the figure's details and colors. When drawing and analysis, according to the relevant line drawing and literature records, we mainly restored the shirt and skin color, and the shape of the crown. The small crown on his head is also called the lotus crown（莲花冠）because it looks like a blooming lotus. The lotus crown was popular in the Tang Dynasty, and it was still popular in later generations. “Zhongdan”（中单 the shirt）is a kind of underwear that is worn in a wide robe, usually made of white Shaluo（纱罗）or Bubo（布帛）. In the Tang Dynasty, the Fangxinquling（方心曲领, square heart curved collar）was a kind of decorative collar attached to the Zhongdan, that is, a semicircular hard lining was set on the Zhongdan to make the collar protrude. In Tang Dynasty, the square heart curved collar was a kind of collar decorated on the official dress for people above Rank seven（七品）, which usually matched with robes, big sleeves coat and Zhongdan, and it was mostly white. The Song Dynasty officers used to wear a square heart curved collar on the cross collar court clothes “following the Tang tradition”, but this kind of ornament in the Song Dynasty was round above and square below, which looks like a keyūra; it is very different from the shape in the Tang Dynasty.

（Text by: Li Yingjun）

图：李迎军 Picture by: Li Yingjun

各国王子听法图服饰
莫高窟初唐第332窟主室
北壁

The prince's clothes of different countries
on the north wall of the main chamber in cave 332
dated to the early Tang Dynasty
at Mogao Grottoes

在唐时长安城流行一句谚语："昆仑奴，新罗婢"，指当时的上流社会流行蓄养昆仑奴仆、新罗婢女的风气。"昆仑奴"是古代到中国的黑人奴仆的泛称，既包括来自南海、印度群岛的卷发黑身的奴隶，也包括来自东非的黑人奴隶。而在各国王子听法图中的昆仑王子，大多认为是南亚、东南亚的海岛民族。

各国王子听法图是维摩诘经变中必不可少的部分，在莫高窟所有各国王子听法图中，几乎都有昆仑王子的形象出现，而且都是位于诸国王子的前列，体貌与着装特征基本一致。

莫高窟第332窟中各国王子听法图中的昆仑王子形象已经斑驳，服装与身体结构只能依稀可辨。壁画上的昆仑王子肤色黝黑，体健壮，高鼻深目、鼻头肥大，嘴唇丰厚，头顶的蓬松发饰已经褪成白色。佩戴圆形耳环、颈圈、臂钏、手镯、脚镯。上身袒裸，缠裹披帛，下身着缠裹式花短裤，结构近似印度传统男裤，赤足。服装以大面积的茜色、深褐色与小面积石青、石绿色组合。仔细辨别各块面料上皆有图案，上身披帛上的图案似是扎经染色纹样。绘制整理这身昆仑王子的形象时，参照了多个版本的《职贡图》中的昆仑奴形象，以及南朝梁画家萧绎所绘《王会图》（也称《番客入朝图》）中描绘的南海狼牙修国使形象，重点借鉴了披帛及下装缠裹方式的表现。

（文：李迎军）

In the Tang Dynasty, there was a popular saying in Chang'an City: "Kunlun Slaves, Xinluo maidservants", which refers to the fashion of having Kunlun slaves and Xinluo maids in the upper class society at that time. "Kunlun slave" is a general term for black slaves in ancient China, including black slaves with curly hairs and black skin from the South China Sea and the Indies, as well as black slaves from East Africa. The prince of Kunlun in the picture of Princes listening Dharma of various countries（各国王子听法图）is mostly believed from South or Southeast Asia island.

The picture of princes listening Dharma of various countries is an essential part of Vimalakirti's Sutra tableau. In all the picture of princes listening Dharma of various countries in Mogao Grottoes, Kunlun Prince appeared almost in all the pictures, and they always stand in the forefront of the group, and their physical appearance and dress characteristics are basically the same.

In cave 332 of Mogao Grottoes, the image of Kunlun Prince in the picture of Princes listening to Dharma of various countries has mottled, the clothes and body can only be roughly distinguished. In the mural, the Kunlun prince has a dark complexion, strong body, deep eyes and high wide nose, thick lips, and the fluffy hair ornaments on his head have faded to white. He wears round earrings, collars, armlets, bracelets and anklets. The upper body is naked, wrapped with uttariya, and the lower body is wrapped with floral shorts. The style is similar to the traditional Indian man's pants, and barefoot. The clothes color are Alizarin red and dark brown which matched with small area of malachite green and azurite. Every part of the fabric has pattern if we look carefully, the uttariya on the upper body seems to be tie-dyeing pattern. When drawing and analys the image of Kunlun prince, we referred to the Kunlun slave image in several versions of *Zhigong Tu*（《职贡图》）, and the image of Langyaxiu（狼牙修国）envoy from the South China Sea depicted in *Wang Hui Tu*（《王会图》）[also known as *the picture of Foreign guest entering the court*（《番客入朝图》）] painted by Liang painter Xiao Yi in the Southern Dynasty. We mainly referred the performance of the uttariya and wrapping shorts.

（Text by: Li Yingjun）

图：李迎军 Picture by: Li Yingjun

敦煌莫高窟初唐

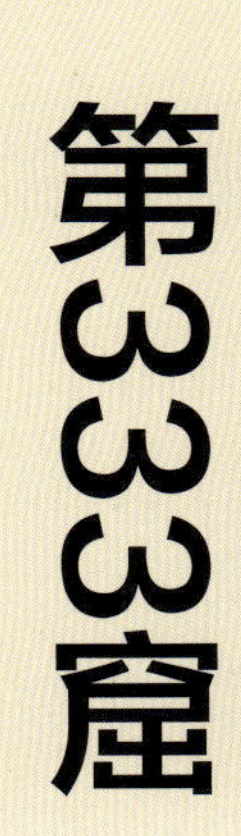

第333窟

Dunhuang Mogao Grottoes Cave 333 of the Early Tang Dynasty

第333窟是初唐时期开凿的覆斗顶小型窟，主室窟顶藻井绘莲花井心，垂角帷幔铺于四披，四披绘千佛。主室西壁设马蹄形佛床，彩塑一佛、二弟子、二菩萨，均经清代重修，床沿存五代绘供养人像。西壁浮塑背光，两侧绘弟子像各三身及赴会佛二组。南北两壁各绘六身供养菩萨和一身弟子像，下部为五代绘供养人像。东壁门上绘千佛，门南北两侧各绘一身弟子像，并残存五代绘供养人像。前室存五代和宋代壁画。

虽然此窟壁画和彩塑的艺术水准在初唐诸多洞窟中并不十分突出，但主室西壁保留了一幅服饰造型明确清晰的弟子像，以及南壁保留的颇具特色的供养菩萨服饰图案。以瓔珞缠绕菩萨长裙使裙身结构更加丰富，并且半透明面料质感的表现非常突出，这在南北两壁供养菩萨服饰上都有所体现，另外菱格纹、小簇花、小团花等多类服饰纹样在此窟中都有所展示。

（文：崔岩）

Cave 333 is a small truncated pyramidal ceiling cave built in the early Tang Dynasty. On the top of the main chamber, the caisson center painted with lotus flower, the fringes of the caisson reach to the four slopes partially, and the four slopes are painted with thousand Buddha motif. On the west wall of the main chamber, there is a horseshoe shaped altar with painted sculptures of one Buddha, two disciples and two Bodhisattvas. All of them were rebuilt in the Qing Dynasty, but the donors painted on the edge of the altar in Five Dynasty. On the west wall, there are backlight in relief, and three painted disciples and two attending Buddhas on each side. There are six Bodhisattvas and one disciple painted on the north and south wall respectively, and the lower area are painted the donors dating back to the Five Dynasties. Thousand Buddha motif is painted above the east gate, both north and south side of the east gate have one painted disciple, along with donors dating back to the Five Dynasty. There are murals of the Five Dynasties and Song Dynasty in the front room.

Although the artistic level of the murals and painted sculptures in this cave is not very prominent among caves in the early Tang Dynasty, the west wall of the main chamber retains a clear image of a disciple, and the south wall retains a distinctive pattern on offering Bodhisattvas' clothes. The long antariya wearing by Bodhisattva wrapped in keyūra makes the structure of the antariya more abundant, and the translucent fabric texture is very prominent, which is reflected in the Bodhisattva clothes both on the north and south walls. In addition, many kinds of clothes patterns such as rhombic pattern, small cluster flowers and small flower medallion are displayed in the cave.

(Text by: Cui Yan)

The disciple's clothes and kasaya pattern in the west niche of the main chamber in cave 333 dated to the early Tang Dynasty at Mogao Grottoes

弟子像服饰、袈裟图案 莫高窟初唐第333窟主室西壁龛内

自十六国开始，佛陀与弟子的题材就在敦煌莫高窟壁画中频繁出现，隋唐以来逐渐形成了以迦叶、阿难为首的十大弟子群像。第333窟中的主尊佛陀与左右两侧的阿难、迦叶彩塑都由于后世的修整和重绘而失去了初唐风貌，所幸主尊彩塑后侧方的这身弟子像壁画虽略斑驳，但还保留着当年绘制的原貌。

画中弟子身体呈直立动势，双手抬于前，左低右高，内着镶有朱红缘边的石绿色僧祇支，外披田相纹袈裟，装饰多色五瓣花纹，盛唐第199窟的高僧形象中也有类似花纹的袈裟形象出现。袈裟为褐色地，上面分别排列着黑色、石绿色和红色的花纹图案。图案造型似梅花，花瓣浑圆，根部细窄，以平涂轮廓式方法表现。从组织结构和艺术风格来看，推测为模板印花工艺制成，具有形象简练、易于重复、色彩单纯等特点，具有平面化的装饰特征。下着黑灰色多褶裙，装饰黑色四瓣散花纹，足蹬红色高齿履。壁画上弟子的肤色已经氧化变深，但依稀可以分辨人物的神态——眉头若蹙、嘴微张，表情刻画细腻传神。

（文：李迎军、崔岩）

Since the Sixteen Kingdoms, the theme of Buddha and disciples has appeared frequently in Dunhuang Mogao Grottoes murals. Then the Sui and Tang Dynasty, ten major disciples headed by Kasyapa and Ananda group images appeared. The main Buddha in cave 333 and the painted sculptures of Ananda and Kasyapa on both sides have lost the original style of the early Tang Dynasty due to the repairing and redrawing of later generations. Fortunately, although the painted disciple on the side behind the main painted sculpture is slightly mottled, it still retains the original appearance.

In the painting, the disciple's body is upright, both hands are raising in front, the left low and the right high. Inside, he wears a malachite green Sankaksika inlaid with vermilion trimming. Outside, he wears kasaya with field pattern, decorated with multi color five petals flower patterns. The kasaya with similar pattern also appears in the image of the eminent monk in cave 199 of the high Tang Dynasty. The kasaya is brown, with black, malachite green and red patterns arranged on it. The pattern is similar to plum blossom, with round petals and narrow ends, which are depicted by flat coloring with outlines. From the perspective of structure and artistic style, it is speculated that this pattern may be made by printing, which has the characteristics of concise image, repetition, simple color and flat decoration. Under the black and gray pleated antariya, decorated by black four petals flower pattern, and the feet wearing red high tooth shoes. In the mural, the skin color of the disciple has been oxidized and darkened, but the facial expression of the characters can be discerned roughly — the eyebrows are frowning, the mouth is slightly open, and the expression is delicate and vivid.

（Text by: Li Yingjun, Cui Yan）

图：李迎军 Picture by: Li Yingjun

图：崔岩 Picture by: Cui Yan

The Offering Bodhisattva's antariya pattern

on the south wall of the main chamber in cave 333 dated to the early Tang Dynasty at Mogao Grottoes

供养菩萨裙身图案

莫高窟初唐第333窟主室南壁

第333窟主室南北壁各绘制六身形态各不相同的供养菩萨，服饰穿搭相似，因此这里选取南壁东向第二身供养菩萨的裙身图案进行整理绘制。这身供养菩萨神情专注，左手持军持，右手擎莲花，侧身面向佛床。上身斜披披帛，束石绿色围腰，下着短裙和半透明长裙，悬垂的璎珞兜住裙侧形成美丽的波浪线。

菩萨皮肤色彩变色严重，所幸裙子的造型和图案保留了原本的形态和色彩。

裙身为浅土红色，上面排列装饰着四方连续的散花纹。花纹保留了自然的对称形态，以四褐色圆点为花心，石绿色花叶呈放射状展开，四个一组进行交错排列。唐代是花鸟植物类图案兴起并占据主流的时期，在自然基础上按照装饰规律进行归纳和整合的散花纹常见于服饰图案中，清新雅致的风格深受人们喜爱。

（文：崔岩）

The north and south walls of cave 333 are painted with six different postures offering Bodhisattvas, and the clothes are similar. The antariya patterns on the second offering Bodhisattva on the south wall from the east is selected for drawing and analysis. This Bodhisattva is dedicated, holding the water vase in his left hand, and holding the lotus in his right hand, and facing the Buddha bed in profile. The upper body is covered by uttariya, the lower body wears malachite green waist wrap, skirt and translucent long antariya. The hanging keyūra holds the antariya side forming beautiful wavy lines.

The color of Bodhisattva's skin changed seriously. Fortunately, the antariya retains its original shape and color.

The antariya is light earth red, and it is decorated with four in a group repeated flower patterns. The pattern retains the natural symmetrical shape, with four brown dots as the flower center and malachite green leaves radiating out in a staggered arrangement. In the Tang Dynasty, flower, bird and plant patterns rose and occupied the mainstream. The dispersive flower patterns were common in the clothes patterns, and the fresh and elegant style was deeply loved by people.

(Text by: Cui Yan)

图：崔岩 Picture by: Cui Yan

敦煌莫高窟初唐

第334窟

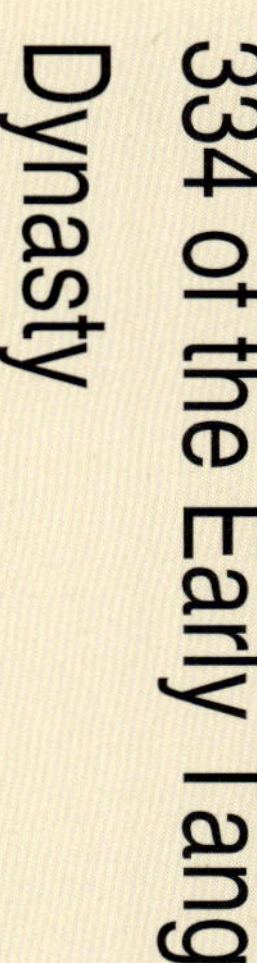

Dunhuang Mogao Grottoes Cave

334 of the Early Tang Dynasty

第334窟是初唐时期开凿的覆斗顶窟。主室窟顶为八瓣莲花藻井，藻井边饰为联珠纹、卷草纹、宝相花等，四披绘千佛。主室西壁开平顶敞口龛，内塑一佛、二弟子、四菩萨，龛外南北两侧各塑一神兽，经清代重修。龛顶绘说法图，龛壁浮塑背光，两侧绘维摩诘经变。龛沿内外侧分别为卷草纹和宝相花纹。龛外南北两侧绘千佛，下部分别绘供养人像二身、侍从四身，整体修长的造型展示了初唐时期的典型服装款式特征。主室南北两壁分别绘弥勒经变和阿弥陀经变，下绘男女供养人像各一排。主室东壁门上绘十一面观音变一铺，两侧绘千佛，下绘供养人及供养牛车、驼马。供养人服饰款式及细节特征都呈现了非常典型的初唐样貌，色彩搭配以大红和石绿为主。

主室西壁龛内壁画保存较好，碧青的虚空中飞云流动、落英缤纷，龛顶中央用五彩祥云围成椭圆形的边框，绘说法图一铺，龛壁绘维摩诘经变，画面清新明朗。此窟壁画内容丰富，保存了较多经典的服饰形象和图案，尤其是服饰的细节保留完好。人物形象丰富各异，其中既有围腰长裤的印度菩萨服装，也有交领长袍的中原服饰。维摩诘经变部分的帝王大臣各国王子的形象并没有如第220窟那般复杂，但是单体形象比较大识别度很好；天女的杂裾垂髾特征十分明确突出，一如前面第332窟。

（文：崔岩）

Cave 334 is a truncated pyramidal ceiling cave excavated in the early Tang Dynasty. On the top of the main chamber is an eight petaled lotus caisson decorated with beads, scrolling grass pattern, Baoxiang patterns on the fringes. The four slopes are covered by Thousand Buddha motif. The west wall of the main chamber has an open niche with flat top; inside the niche, there are one Buddha, two disciples and four Bodhisattvas painted sculptures, and outside the niche. There are two sacred animals on the north and south sides, all repainted in the Qing Dynasty. The top of the niche is painted with a preaching scene, and the niche wall has backlight relief, and the two sides are painted with Vimalakirti Sutra tableau. Along the inside and outside of the niche are scrolling grass pattern and Baoxiang pattern. The south and north sides of the niche are painted with Thousand Buddha motif, and the lower parts are painted with two donors and four attendants respectively. The overall slender figures show the typical clothes style characteristics of the early Tang Dynasty. The north and south walls of the main chamber are painted with Maitreya Sutra tableau and Amitabha Sutra tableau respectively, and the lower parts are painted with a row of male donors on one side and a row of female donors on the another side. Above the east door of the main chamber has a painted 11 heads Guanyin, Thousand Buddha motif on both sides, donors, ox carts and camels, horses on the bottom area. The clothes styles and details of the donors present a very typical appearance of the Tang Dynasty, colors are mainly malachite green and azurite.

The murals in the west niche of the main chamber are well preserved, with flying clouds flowing and falling in the blue void. The center of the niche top is surrounded by colorful auspicious clouds to form an oval frame, and the picture is fresh and clear. The murals in this cave are rich in content, and many classic clothes images and patterns are preserved, especially the details of clothes. The images of the characters are various, including Indian Bodhisattva clothes with girdle trousers and Central Plains clothes with cross collar robes. The images of emperors, ministers and princes in the Vimalakirti Sutra tableau are not as complicated as those in cave 220, but the single image is relatively large and easier to discern; the features of celestial woman's Zajuchuishao（杂裾垂髾）is very clear and prominent, just as in cave 332.

（Text by: Cui Yan）

Hua Bodhisattva's clothes

in the west niche of the main chamber in cave 334 dated to the early Tang Dynasty at Mogao Grottoes

化菩萨服饰

莫高窟初唐第334窟主室西壁龛内

这是第334窟主室西壁龛内维摩诘经变中的化菩萨，表现的是“香积佛品”的故事情节。因文殊菩萨与维摩诘辩论，时间已晚，舍利弗担心与会众人没有饭食。于是，维摩诘以神通力作化菩萨，向众香国香积如来借来满钵香饭，令与会菩萨天人众皆得饱腹、身安快乐。从壁画中可以看到化菩萨神态和悦，正双手举钵向维摩诘供奉，所以又被称为“香积菩萨”。

维摩诘所说经中说化菩萨“相好光明，威德殊胜”，因此，壁画也着重表现了她优美的身姿和飘逸的服饰。菩萨姿态优雅，单腿跪于覆瓣莲花座上，是初唐出现的新题材和新的表现形式。她头戴镶嵌宝珠的仰月冠，束发的长带绕手臂下垂飘荡。身上佩戴璎珞、臂钏和手镯，腰带为方块式嵌宝珠的金属质地，精致华丽。服饰造型上，菩萨上身袒裸，披轻薄透明的天衣，下着深青色短裙和透明长裙，裙缘为鲜艳的石绿色，随着菩萨的身躯起伏形成美丽的衣褶。画面艺术处理上，线条轻盈灵动，色彩斑斓，错落有序，呈现了菩萨下凡、普惠信众的浓郁氛围。

（文：刘元凤）

This is the Hua Bodhisattva（化菩萨）in the Vimalakirti Sutra tableau in the west niche of the main chamber of cave 334, which shows the story the chapter of Xiangji Buddha（香积佛品）. Because the time was very late when Manjusri debated with Vimalakirti, Sariputra worried that there would be no food for the participants. As a result, Vimalakirti used his divine power as a Bodhisattva to borrow a bowl full of delicious food from Xiangji Tathagata（香积如来）, so that all participating Bodhisattvas and celestial beings could be fed and feeling comfortable. From the murals, we can see that Hua Bodhisattva is in a kind and happy manner, holding a bowl with both hands to worship Vimalakirti, so it is also known as Xiangji Bodhisattva（香积菩萨）.

In the Vimalakirti Sutra, it is said that the Hua Bodhisattva is “good looking and shining, elegant and noble”. Therefore, the mural also focused on his beautiful posture and elegant clothes. The graceful posture of Bodhisattva, kneeling on the lotus seat with one leg, is a new theme and new form of expression in the early Tang Dynasty. He wears the moon crown inlaid with jewels, and the long band tied up hairs hung around his arms. He wears keyūra, armlets and bracelets. The belt is made of metal in square shaped and inlaid jewels, which is exquisite and gorgeous. In terms of dressing style, the Bodhisattva's upper body is naked, wearing a piece of light and transparent heavenly cloth, and a dark blue short antariya and a transparent long antariya. The antariya trimming color is bright malachite green, forming beautiful pleats along with the Bodhisattva's body. In terms of artistic treatment, the lines are light and flexible; the colors are colorful and orderly, presenting a strong atmosphere of Bodhisattva coming down to earth and benefiting all believers.

（Text by: Liu Yuanfeng）

图：刘元风 Picture by: Liu Yuanfeng

The celestial woman's clothes

in the west niche of the main chamber in cave 334 dated to the early Tang Dynasty at Mogao Grottoes

天女服饰

莫高窟初唐第334窟主室西壁龛内北侧

这是初唐第334窟主室西壁龛内北侧维摩诘经变“观众生品”中的天女，其面部神态洒脱。据维摩诘所说经中所载，此天女散花戏弄舍利弗，以花朵的落与不落比喻修行中的分别之想。

天女神态平和，昂首行进，左手执羽扇，右手托花苞。头戴高耸的双凤冠，鬓边插如意云头钗，鬓发卷曲。身着红色的半袖衫，领部为石绿色的曲折翻领结构，袖部有蓬松的石青色荷叶边装饰。内穿深青色大袖襦，袖缘为石青色。以腰带束围腰和蔽膝，长裙曳地，裙边装饰有荷叶边，裙侧华袿飞扬。天女所服襳髾，又称“杂裾垂髾”，是在围腰下施加相连的三角形飘带装饰，走动时衣带当风、如燕飞舞，《周礼正义》形容说：“其下垂者，上广下狭，如刀圭。”因为三角形装饰如刀圭状，所以又名“袿衣”，宽松大度的服饰与纷飞的天花交相辉映。画面艺术处理上，用笔奔放而劲健，色彩配置对比强烈而醒目。在服饰的整体造型上，还保留有魏晋格调，呈现其贵族女性的雍容大气之风范。

（文：刘元风）

This is the celestial woman from Vimalakirti's Sutra tableau, the chapter of observing all sentient beings（观众生品）in the west niche of the main chamber of cave 334 dating back to the early Tang Dynasty, her face is free and easy. According to Vimalakirti Sutra, this celestial woman teased Sariputra by scattering flowers, using the falling and not falling flowers as a simile to explain the thought of differentiation.

The celestial woman has a peaceful expression, head up, holding a feather fan in her left hand and a flower bud in her right hand. She wears a high double phoenix crown on the head, a wishful cloud hairpin in hairs, and curly hairs cover her temples. She wears a red half sleeves shirt, the collar is a malachite green tortuous lapel structure, and the sleeves are decorated with fluffy azurite ruffles. The inside of the sleeves are dark blue shirt with big sleeves, and the sleeves' trimming are azurite. The belt holds the waist wrap and Bixi, the long antariya touching the ground, and the hem is decorated with ruffles. This celestial woman's clothes called as Xianshao（襳髾）, also known as the Zajuchuishao（杂裾垂髾）, is a kind of clothes has triangular ribbons below the waist. When she walks, the ribbons are like swallow flying around. *Zhouli Zhengyi*（《周礼正义》）describes: “the hangings are wide in the upper part and narrow in the lower part, just like a knife.” Because the triangle decoration is like a knife, it's also known as Guiyi（袿衣）. The loose and comfort clothes and the flying flowers complement each other. In the art analysis, the lines are bold and vigorous, and the color configuration is strong and eye-catching. In the overall shape of the dress, it also retains the style of the Wei and Jin Dynasties, showing the elegant style of noble women.

（Text by: Liu Yuanfeng）

图：刘元风 Picture by: Liu Yuanfeng

Bhiksu's clothes
in the west niche of cave 334 dated to the early Tang Dynasty at Mogao Grottoes

比丘像服饰
莫高窟初唐第334窟西壁龛内

第334窟西壁龛内两侧绘有维摩诘经变壁画，南北两侧各绘有神采飞扬的维摩诘像，身边随诸多听法圣众。整幅壁画绘于统一的石青底色上，天地连属，如住虚空，澄明清净。这里选取西壁龛内北侧下部的舍利弗形象进行整理绘制。舍利弗在《佛说维摩诘经》中多次发问，通过与维摩诘的对话引出要阐发的深意，是经文中的重要人物之一。在经文《不思议品·第六》中写道，舍利弗看到维摩诘所处居室中没有床座，担心诸菩萨大弟子众没有座位，于是维摩诘借此缘故，显示神通力，由须弥灯王如来送来三万二千师（狮）子座，由此演说不可思议解脱的法门。因此，壁画中就绘制了舍利弗跏趺坐于狮子座上的形象，狮子座即为胡床的样貌。

壁画中的舍利弗呈比丘相，内穿茜色覆头衣，外穿石绿色通肩袈裟，领口与肩部自然翻折出袈裟内里，衣口自然下垂。舍利弗的神情庄重虔诚，双目圆睁，聆听维摩诘说法，面部形象写实，表现手法简洁。绘制整理时侧重于舍利弗所披袈裟的垂感表现与坐于狮子座上而形成的自然褶皱。

（文：吴波）

In the west niche of cave 334, the two sides are painted with Vimalakirti Sutra tableau, and the north and south sides are painted with Vimalakirti images respectively, accompanied by many followers. The whole picture uses azurite as background, so the heaven and the earth are connected, just like living in void, which is clear and pure. This paper selects the image of Sariputra in the lower part of the north side of the west niche to analyse. In the *Buddhist scripture of Vimalakirti sutra*, Sariputra asked questions many times. Through the dialogue with Vimalakirti, Sariputra elicited the profound meaning need to be elucidated. He is one of the most important figures in the Sutra. In the *Incredible chapter No.6*《不思议品·第六》, it is described that Sariputra saw that there was no seat in the room where Vimalakirti lived, and worried that the disciples and Bodhisattvas have no place to sit. Therefore, Vimalakirti showed his supernatural power, asked Xumidengwang Tathagata（须弥灯王如来）to send 32000 lion seats here. By doing this, he made a speech of the miraculous method to liberation. Therefore, the mural depicts the image of Sariputra sitting on Lion seat, which looks like Hu bed.

In the mural, Sariputra looks like a bhikkhu. He wears a Rubi colored sweater inside with head covered, and a malachite green kasaya with both shoulders wrapped. The neckline and shoulders of the kasaya naturally fold out of the lining, and the edges of the kasaya naturally droops. Sariputra's expression is solemn and pious; his eyes are wide open, listening to Vimalakirti's statement. His face is depicted realistically, simple and clean. When drawing and studying, we focused on the soft textile of Sariputra's kasaya and the natural folds formed by sitting on the lion seat.

（Text by: Wu Bo）

图：吴波 Picture by: Wu Bo

The emperor's clothes in the west niche of cave 334 dated to the early Tang Dynasty at Mogao Grottoes

帝王听法图服饰 莫高窟初唐第334窟西壁龛内

第334窟西壁龛内听法众之一，服装形制类似帝王，腰带附有蔽膝，垂下大绶，广袖袍服交领、左衽。

段文杰先生在《敦煌石窟艺术研究》一书中提到莫高窟第334窟中维摩诘变中的听法帝王形象："戴通天冠，着白单衣、曲领、大带、大绶、蔽膝、分稍履。"并且衣袖宽大，未露手，这在唐代同时期的画作中也可见到类似情形。传阎立本作《历代帝王图》中的帝王多未露手，莫高窟第220窟洞壁"维摩变"中的听法帝王和盛唐第103窟中的听法帝王也未露手。参照唐代其他洞窟中的帝王蔽膝，推测第334窟帝王蔽膝应有图案，只是经年累月，模糊不清而已。此外双肩应有日月纹样，现已无法辨识。对于左衽这个问题，参考《历代帝王图》中对于帝王形象的表现，均为右衽或对襟，莫高窟第220窟和103窟中的帝王听法形象为对襟，同时期出土文物中的交领也基本为右衽，左衽在历史上一度被认为是蛮夷之地的穿着方式。然而本图尊重壁画原貌，其着装为左衽。

（文：张春佳）

It is one of the audience listening the dharma in the west niche of cave 334. His clothes is similar to emperor's robe, which the belt holds Bixi, a large band hanging down, a wide sleeves robe with cross collar, right side covering the left side.

Mr. Duan Wenjie mentioned the King image in the Vimalakirti sutra tableau in cave 334 of Mogao Grottoes in his book *Research on Dunhuang Grottoes Art*（《敦煌石窟艺术研究》）: "wearing the heaven crown, white thin clothes, round collar, big belt, big band, Bixi and split ends shoes." Moreover, the sleeves are wide and the hands are not exposed, which can be seen in the paintings of the same period in the Tang Dynasty. It is said that most of the emperors hands in Yan Liben's *Painting of Emperors of All Dynasties*（《历代帝王图》）have not been exposed, and the king in cave 220 of Mogao Grottoes and the king in cave 103 of the high Tang Dynasty are also with their hands hidden. Referring to the emperor's Bixi in other caves of the Tang Dynasty, it is speculated that the emperor's Bixi in cave 334 should have patterns, which had faded over the years. In addition, there should be sun and moon patterns on the shoulders, which can't be recognized now. As for the left lapel, referring to the images of emperors in the picture of Emperors of All Dynasties, they are all right lapels or parallel lapels. The images of emperor listening to Dharma in cave 220 and 103 of Mogao Grottoes are parallel lapels. At the same time, lapels style in unearthed cultural relics is also basically right lapels. Left lapels were once considered as the way of barbarian custom in history. However, this drawing respects the original appearance in the mural, and its dress is left lapels.

(Text by: Zhang Chunjia)

图：张春佳 Picture by: Zhang Chunjia

维摩诘服饰

莫高窟初唐第334窟西壁龛内

Vimalakirti's clothes

in the west niche of cave 334 dated to the early Tang Dynasty at Mogao Grottoes

敦煌唐代壁画颇注意人物之间的关系，彼此呼应，互相烘托，整个画面形成人物间有机的统一体。第334窟西壁龛内主尊彩塑两侧分别绘有维摩诘像，以及比丘等听法众。两身维摩诘像所穿服饰形制基本一致，虽然壁画中肤色部分因氧化已看不出原来的设色，但服饰的结构基本清晰，都是外披开襟短袖袍服，头戴两侧垂有长飘带的帽式，脚穿高齿履。不同之处在于北侧的维摩诘内穿左衽广袖深衣，南侧的维摩诘内着深色曲领中单。维摩诘所披短袖外袍与所戴帽式在敦煌初唐画迹中比较少见，帽式从形态上不似传统的纶巾，在敦煌研究院早期的临摹作品中表现出清晰的帽子结构特征，在服饰史中未查到相应的专业名称，可待研究。维摩诘的形象在敦煌壁画中多有表现，一般都是以坐姿讲经论道的形态居多。难得的是第334窟的维摩诘是以站姿出现，可以展示其着装的全貌。在维摩诘服饰效果图整理时基本遵从原有的动态与服饰结构，两尊维摩诘像形态生动，仿佛在怡然自得地讲经论道，一尊维摩诘目光炯炯，深思飞扬，右手放在胸前，似在专心聆听。另一尊维摩诘举止庄重，神态自如，两手从外袍中伸出，似在侃侃而谈，颇具唐朝文人士大夫的风貌。

（文：吴波）

Dunhuang murals in the Tang Dynasty payed attention to the relationship between the characters, echoing with each other, contrast with each other, the whole picture forms an harmonious unity among the characters. In the west niche of cave 334, the two sides of the main painted sculpture are painted with Vimalakirti and bhikkhu. The shape and color of the clothes worn by the two Vimalakirti paintings are basically the same. Although the original color of the skin part in the mural can not be seen due to oxidation, the structure of the clothes is basically clear. They are all covered with open-lapels short sleeves robes, wearing hats with long bans on both sides, and wearing high toothed shoes. The difference is that the Vimalakirti on the north side wears a left lapel wide sleeve dart shirt, and the Vimalakirti on the south side wears a dark curved collar middle shirt. The short sleeves robe and the hat style of Vimalakirti are rare in the early Tang Dynasty paintings in Dunhuang. The hat style is not similar to the traditional Guanjin [纶巾, silk head dress resembling ridged roof (for man)] in shape. The clear hat structure characteristics can be found in the early copying works of Dunhuang Research Academy. We haven't found corresponding professional name for the hat in the history of clothes, which needs to be studied more. The image of Vimalakirti are common in Dunhuang murals, generally in the form of sitting, preaching and debating. What is rare is that Vimalakirti in cave 334 is depicted in a standing posture, which the whole picture of his clothes can be seen. When doing the effect drawing of Vimalakirti's clothes, we basically follow the original dynamics and clothes structure. Two Vimalakirti's images are vivid in shape, as if they were happy to preach the scriptures. One Vimalakirti's eyes are bright, thinking deeply, and his right hand on his chest, as if he is listening attentively. Another Vimalakirti has a dignified manner and a free expression. His hands stretch out from his robe and seem to be talking freely. This two images have the style of literati and officers in the Tang Dynasty.

(Text by: Wu Bo)

图：吴波 Picture by: Wu Bo

供养菩萨服饰

莫高窟初唐第334窟主室东壁门上南侧

The offering Bodhisattva's clothes

on the south side of the east wall above the door of the main chamber in cave 334 dated to the early Tang Dynasty at Mogao Grottoes

初唐第334窟的这尊菩萨是东壁窟门上方十一面观音菩萨莲花宝座两侧的供养菩萨之一，由于绘在背光墙面避免了强光的侵蚀，整体形象得以清晰、完整地保存，人物形态、服装结构都明晰可辨，色彩也艳丽如新。

在十一面观音菩萨南侧的莲台上，这尊供养菩萨目视前方、神情坦然，默默地向佛祈祷。她单膝跪地，辅以双手合十的动作，充分体现出对佛法的无限虔诚与敬仰。这尊菩萨肤色白皙，体型已初现唐朝时期的丰腴之态，脑后的披肩卷发还保留着西域人物的形象特征，头顶盘圆髻，戴仰月冠，上身斜披披帛，颈上戴颈环及两条联珠式璎珞项链，两支手臂与手腕均佩戴臂钏与手镯，下着红色点花纹罗裙，系绿色围腰。绘制整理的图像如实地表现了这尊供养菩萨的造型特征与服饰结构，在保持服饰红绿混搭的配色关系与色彩面积的同时，适度降低了色彩的纯度，以凸显整体形象的古拙之风。

（文：李迎军）

This offering Bodhisattva in cave 334 dating back to the early Tang Dynasty is painted above the east door, beside the 11 headed Avalokitesvara Bodhisattva, close to the lotus throne. Because it is painted on the dark side, it avoids the erosion of strong light; the whole image is clearly and completely preserved. The figure shape and clothes structure are clear and recognizable, and the color is as bright as new.

At the south side of the 11 headed Avalokitesvara Bodhisattva, this offering Bodhisattva is siting on lotus seat, looking straight ahead, peaceful, and praying to the Buddha silently. He kneels on one knee, with his hands folded, which fully reflects his infinite piety and respect for Dharma. This Bodhisattva has a white complexion and the figure is plump which is already very close to the Tang Dynasty style. His curly long hair at the back of his head still retain the characteristics of Western Regions people. His hairs tied up as a bun, a crescent moon and sun crown on his head, a uttariya on his upper body, a collar and two beads keyūra on his neck, two arms and two wrists wear armlets and bracelets, and a green waist wrap, red antariya with dotted flower pattern. The drawing shows the modeling characteristics and clothes structure of this Bodhisattva. While maintaining the color matching relationship and color area of red and green, the purity of colors are moderately reduced to highlight the ancient style of the overall image.

(Text by: Li Yingjun)

图：李迎军 Picture by: Li Yingjun

彩塑菩萨裙身图案

莫高窟初唐第334窟主室西壁龛内

The painted sculpture Bodhisattva's antariya pattern

in the west niche of the main chamber in cave 334 dated to the early Tang Dynasty at Mogao Grottoes

第334窟主室西壁龛内塑有四身菩萨，菩萨上身的肌肤和服饰均经清代重修，但菩萨的裙子仍然保留了初唐的原貌，且绘制精美，因而选取其中两身菩萨的裙饰图案进行整理绘制。

菩萨所穿长裙以红色为底，以四方连续的石绿色、石青色散花纹、卷云纹为花，然后依据身体部位的相应装饰位置，以条带方式绘出不同题材和色彩的二方连续图案。竖向图案装饰于股前，横向图案装饰于膝下裙缘处，这是一种将图案装饰与服装穿着巧妙结合的定位设计方法，同样的例子在初唐第328窟彩塑菩萨裙饰图案中也可以看到。

（文：崔岩）

There are four painted sculptures Bodhisattvas in the west niche of cave 334. The upper body skin and clothes of Bodhisattvas have been repainted in the Qing Dynasty, but the antariyas still retains the original appearance of the early Tang Dynasty which is very exquisite. Therefore, the antariya patterns of two Bodhisattvas are selected for analysis and studying.

This two Bodhisattvas wear long antariyas with red as the background, four in a group repeated malachite green, azurite dispersive flower pattern, rolled cloud pattern as decoration, and then drew two as a group repeated stripe patterns of different themes and colors according to the corresponding parts of body. The vertical pattern is decorated in front of the thigh, and the horizontal pattern is decorated at the antariya trimming under the knee. This is a positioning design method that skillfully combines pattern decoration with clothes. The same example can be seen in the pattern of painted sculpture Bodhisattva antariya in cave 328 of the early Tang Dynasty.

(Text by: Cui Yan)

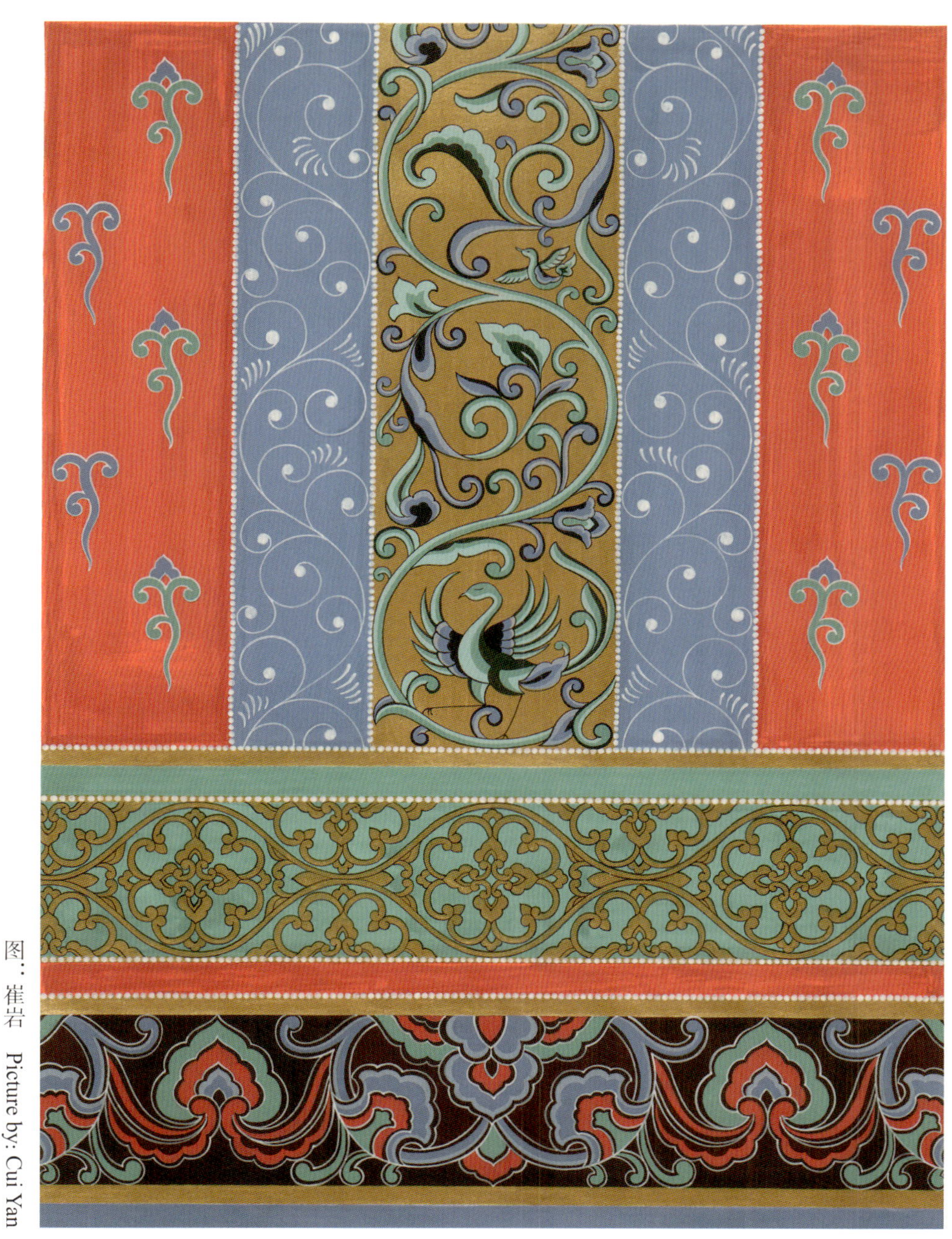

图：崔岩 Picture by: Cui Yan

The painted sculpture Bodhisattva's antariya pattern

in the west niche of the main chamber in cave 334 dated to the early Tang Dynasty at Mogao Grottoes

彩塑菩萨裙身图案

莫高窟初唐第334窟主室西壁龛内

菩萨裙饰图案的题材十分丰富，包括卷草纹、卷草异兽纹、宝相花纹等，色彩以朱红、石青、石绿为主，有的施以平涂，有的以退晕法表现，对比强烈，缤纷艳丽。白色和金色在图案中起到重要的调节作用，令画面色彩更加和谐统一。

虽然此窟彩塑菩萨的裙饰图案保存较好，但随着时间的流逝，也出现了不同程度的褪色、变色等现象，因此，作者参考了常沙娜老师在《中国敦煌历代服饰图案》一书中绘制于20世纪50年代的整理图稿，以求追溯更加贴近真实、绚烂华丽的初唐服饰图案面貌。

（文：崔岩）

The motifs of Bodhisattva antariya patterns are very rich, including scrolling grass pattern, scrolling grass pattern with different animals, Baoxiang pattern, etc. The colors are mainly vermilion, malachite green and azurite. Some are flat coloring, and some are expressed by color-gradation technique. The contrast is strong and the colors are bright and gorgeous; white and gold play an important role in the pattern, making the color of the picture more harmonious and unified.

Although the patterns of the painted Bodhisattvas' antariyas in this cave are well preserved, as time flies, the colors also faded and changed in varying degrees. Therefore, the author refers to the drawings drawn in the 1950s by Chang Shana in his book *Decorative Design from China Dunhuang Mural* (《中国敦煌历代服饰图案》), in order to trace the more realistic and gorgeous clothes patterns of the early Tang Dynasty.

(Text by: Cui Yan)

图：崔岩 Picture by: Cui Yan

彩塑阿难尊者偏衫图案

莫高窟初唐第334窟主室西壁龛内

The pattern on painted sculpture Ananda's Sankaksika

in the west niche of the main chamber in cave 334 dated to the early Tang Dynasty at Mogao Grottoes

第334窟主室西壁龛内彩塑阿难尊者的头部和肌肤经清代重修，失去原貌，所幸服饰仍保留了初唐风范。阿难为佛陀十大弟子之一，善记忆，对于佛陀之说法多能记诵，故誉为“多闻第一”。敦煌壁画和彩塑中常常把这位弟子塑造成年轻英俊的少年形象，其服装也具有较多装饰和变化。此窟中的阿难尊者身披袒右田相纹袈裟，内着偏衫，下着裙。服装的大面积色彩为红色和石绿色，形成鲜明对比，这里选取阿难所着偏衫图案进行整理绘制。

阿难尊者的偏衫图案为缠枝葡萄纹，主题明确突出。据《汉书·西域传·大宛国》记载：“汉使采蒲陶、目宿种归。”说明汉使从西域得此物种，之后在中原普遍种植。葡萄这种来自西域的特色植物作为织物主题纹样出现在纺织品上，始于东汉时期，至唐代运用更加广泛，体现了唐朝社会的审美趋向与对外交流发展。图案采用复杂的缠枝结构，以藤蔓为骨架，其间填充葡萄与葡萄叶，使图案既饱满又流畅。色彩则采用简洁的配置，以褐色为地，以石绿色为花，沿着纹样边缘进行勾线的处理方式，以及复杂的缠枝结构，使得图案并不单调，反而展现了独属于初唐的勃勃生机。

（文：崔岩、梁霄）

The head and skin of the venerable Ananda painted sculpture in the west niche of cave 334 have been repainted in the Qing Dynasty and have lost their original appearance. Fortunately, the clothes still retain the early Tang Dynasty style. Ananda was one of the major ten disciples of Buddha. He is good at memorizing and can recite many Buddha's teachings, so he is known as "the most hearer". In Dunhuang murals and painted sculptures, this disciple is often portrayed as a young and handsome young man, and his clothes also have many decorations and changes. In this cave, the venerable Ananda wears field pattern kasaya with right shoulder not covered, and a shirt inside and dhoti on the lower body. The main colors of the clothes are red and malachite green, forming a sharp contrast. Here, we select the pattern of Ananda's shirt for studying and drawing.

Ananda's shirt is decorated by twisted vine grape pattern（缠枝葡萄纹）and the theme is clear and prominent. According to the records in the book of *Han, introduction of the Western Regions, Dayuan state*《汉书·西域传·大宛国》, "the Han envoys collected grape and alfalfa seeds back to China." This indicates that the Han envoys obtained this species from the Western Regions and then planted it in the Central Plains. Grapes, as a kind of new plant brought from the Western Regions, appeared on textile as a theme pattern in the Eastern Han Dynasty, which was more widely used in the Tang Dynasty, reflecting the aesthetic trend of the Tang Dynasty and the development of communications with foreign countries. The pattern adopts a complicated twisted vine structure, and uses vines as the frame; it is filled with grapes and grape leaves, making the pattern full and smooth. The color is simple, with brown as the background, malachite green as the pattern, along the the pattern added outlines, including the complex structure, which makes the pattern not monotonous, instead it shows the vitality of the early Tang Dynasty.

（Text by: Cui Yan, Liang Xiao）

图：崔岩 Picture by: Cui Yan

Dunhuang Mogao Grottoes Cave

335 of the Early Tang Dynasty

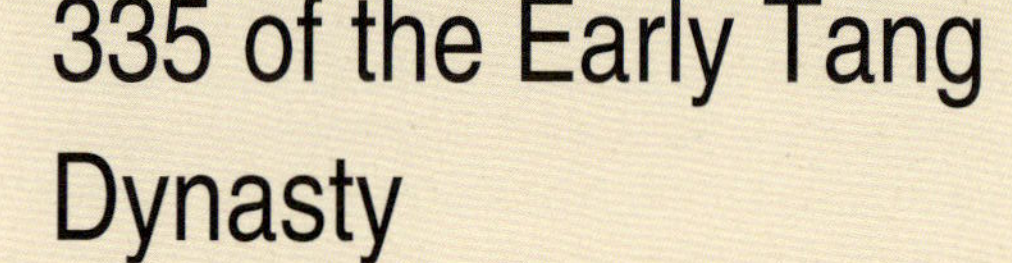

敦煌莫高窟初唐

第335窟

第335窟是初唐时期开凿的覆斗顶窟。主室窟顶为牡丹团花纹井心，卷草、垂幔铺于四披，四披绘千佛。主室西壁开平顶敞口龛，龛内唐塑一佛、一弟子，清塑一弟子、四菩萨。龛顶绘法华经变，龛壁浮塑背光两侧绘法华经变与劳度叉斗圣变。龛外两侧绘观世音菩萨、大势至菩萨各一身。主室南壁绘阿弥陀经变，北壁绘维摩诘经变。东壁门上绘阿弥陀佛一铺，门南绘说法图五铺、药师佛一身、千佛一部，门北存中唐绘说法图一铺，经宋代重描，门两侧下存元代绘菩萨。前室西披残存观音变三铺，其他壁面和甬道存中唐和宋代壁画。

此窟有“垂拱二年”（686年）、“圣历年间”（698～699年）、“长安二年”（702年）等三处题记可以帮助判断开凿和修建时间。洞窟的主题壁画是阿弥陀经变和维摩诘经变，均与净土思想有关。敦煌莫高窟的维摩诘经变到初唐时期趋于成熟，表现内容越来越丰富，此窟北壁通壁绘维摩诘经变，论辩双方同绘于一壁而无分割，这在唐代同类题材中是规模最大且构图形式最完整和特殊的一幅。从这幅壁画下部“方便品”表现的国王、大臣、长者、居士、婆罗门、诸王子等众多人物形象中，能够获取造型各异、内涵丰富、地域特色鲜明的服饰资料，整体造型均为修长之态。可惜画面变色严重，有些服饰细节不易辨识，因此采用文献考证和图像分析的方法进行表现。此窟壁画所绘人物体态和面相较为健壮，同时注重服饰对人物性格的塑造和故事情节发展的推动作用。最为典型的是主室西壁龛内劳度叉斗圣变中乳圆体丰的外道信女，不仅体现出唐代崇尚丰满健康的审美趣味，而且也通过其服饰动态表现成为体现故事结局和趣味的重要载体。

（文：崔岩）

Cave 335 is the truncated pyramidal ceiling cave excavated in the early Tang Dynasty. At the top of the main chamber is caisson with peony pattern in the center. The scrolling grass pattern and the fringes reach to the four slopes. The four slopes are painted with thousand Buddha motif. The west wall of the main chamber has an open niche with a flat top; in the niche, there are painted sculptures of one Buddha and one disciple dating back to the Tang Dynasty, one disciple and four Bodhisattvas dating back to Qing Dynasty. The top of the niche is painted with Lotus Sutra tableau, and the niche wall has backlight relief, painted with Lotus Sutra tableau and the illustration of Raudraksa's Battle with Sariputra（劳度叉斗圣变）. The outer sides of the niche are painted with Avalokitesvara Bodhisattva and Mahasthamaprapta Bodhisattva respectively. The south wall of the main chamber is painted with Amitabha Sutra tableau, and the north wall is painted with Vimalakirti Sutra tableau. A picture of Amitabha Buddha is painted above the east gate, five preaching scenes, a picture of Bhaisajyaguru, and thousand Buddha motifs are painted on the south side of the gate, and a picture of preaching scene painted in the middle Tang Dynasty on the north gate, but it was redrawn in the Song Dynasty. There are painted Bodhisattvas in the Yuan Dynasty on both lower sides of the gate. On the west slope of the front room remain three Avalokitesvara Bodhisattva tableau. On other walls and corridors, there are murals of the middle Tang Dynasty and Song Dynasty.

There are three inscriptions in this cave, which can help us to speculate the time of excavation and construction. The main mural in this cave are Amitabha sutra tableau and Vimalakirti sutra tableau, both of which are related to the idea of pure land. The Vimalakirti Sutra in Dunhuang Mogao Grottoes became more and more mature in the early Tang Dynasty, and its content became more and more abundant. The north wall of this cave is painted with Vimalakirti Sutra tableau, and the whole sutra is on one wall without division. This is the largest and most complete and special composition of similar themes in the Tang Dynasty. From the images of kings, ministers, elders, residents, Brahmins, princes and many other characters in the "the chapter of convenience（方便品）" at the bottom of the mural, we can get a lot of information about clothes, rich connotations and distinctive regional characteristics. The overall figures are slender, unfortunately, the discoloration and oxidation are serious, and some clothes details are not easy to identify, so the methods of literature research and image analysis are used to study. The body and face of the characters in the murals are relatively strong, at the same time, the craftsman payed attention to the clothes in shaping the character and promoting the development of the plot. The most typical is the non-Buddhist nun in the illustration of Raudraksa's Battle with Sariputra in the west niche of the main chamber, who has big breasts and wide hips. It not only embodies the aesthetic taste of advocating fullness and health trend in the Tang Dynasty, but also makes it an important carrier of the story ending and interest through the dynamic performance of their clothes.

（Text by: Cui Yan）

The non-Buddhist woman's clothes

on the south side of the west niche in cave 335 dated to the early Tang Dynasty at Mogao Grottoes

外道女服饰

莫高窟初唐第335窟西壁龛内南侧

莫高窟第335窟龛内南壁的风吹外道魔女，是劳度叉斗圣变中的人物形象。劳度叉斗圣变又称“祇园记图”，在敦煌壁画中最早出现在北魏时期，大量出现于晚唐后，初唐仅此一幅。原文来自《贤愚经·须达起精舍品》，后经变文发展演绎，故事情节越来越丰富。此处画面表现的是劳度叉与舍利弗斗法时其中一个回合的情景——劳度叉化作参天大树，“时舍利弗，便以神力，作旋岚风，吹拔树根，倒著于地，碎为微尘。”最后舍利弗获全胜，六师外道悉归佛法。风吹外道魔女就是这一情节中的一个细节表现：外道魔女正在跪着搔首弄姿地施展本领意欲诱惑舍利弗之时，突然狂风大作，魔女衣带纷飞，惊慌失措。

壁画中盘坐于劳度叉帷帐下方地上的外道魔女乳圆体丰，继承了印度造像艺术中对女性体征的描绘传统。她所穿圆领紧身短襦与下装上的五瓣小花纹样轻巧细致，披巾被大风从肩头吹落飞扬在身后，风动的姿态极富张力。由于绘画颜料的氧化变色，部分服饰的结构、颜色以及下装的形态已经无法准确分辨。绘制整理时，参照敦煌研究院李其琼先生的临绘图稿画清了袖口的双层造型及下装的线条结构。

（文：李迎军）

The wind blowing witch on the south wall of cave 335 in Mogao Grottoes is the character in the illustration of Raudraksa's Battle with Sariputra. Raudraksa's Battle with Sariputra, also known as “Jetavanavihara story illustration（祇园记图）”, first appeared in the Northern Wei Dynasty in Dunhuang murals, a large number of which appeared after the late Tang Dynasty, only this one in the early Tang Dynasty. The original text comes from *Damamūka-nidāna-sūtra·Sudatta build vihara*（《贤愚经·须达起精舍品》）. After the development of Bianwen（变文）, the plot became more and more abundant. The picture here shows the scene of one of the rounds in the fight between Raudraksa and Sariputra Raudraksa turned into a tall tree, “Sariputra, he used the divine power to spin the wind, pulled up the roots, the tree then fell to the ground and broke into dust.” In the end, Sariputra won, and the six masters converted to Buddhism. The wind blowing on the witch is one of the details of this plot: the witch is kneeling, scratching her head and posturing to show her beauty to seduce Sariputra; suddenly the wind blow violently, and the witch's clothes were flying and she was in panic.

In the mural, the witch is sitting on the ground under the tent of Raudraksa; she has big breasts and wide hips, which inherits the tradition of depicting female figures in Indian sculpture art. She wears a round neck, tight short shirt, and five petals small flowers pattern on the pants, the pattern is light and delicate. Her shawl was blown off her shoulders and flying behind by the strong wind. Her wind-driven posture was full of tension. Due to the oxidation and discoloration, the structure, color and some clothes can not be accurately distinguished. In the process of drawing, the double-layer design of the cuffs and the line structure of the clothes on lower body were drawn by referring to the drawing by Mr. Li qiqiong of Dunhuang Research Academy.

（Text by: Li Yingjun）

图：李迎军　Picture by: Li Yingjun

听法随侍大臣服饰

莫高窟初唐第335窟主室北壁

The ministers' clothes

on the north wall of the main chamber in cave 335 dated to the early Tang Dynasty at Mogao Grottoes

在第335窟主室北壁维摩诘经变中，与各国王子同列的有两位唐朝的随侍大臣。二人相向而立，均手持卷子，虽然无法分辨清楚五官神态，但从身体语言判断二人似在相互交流。两位官员均穿方心曲领中单，外罩大袖袍服。他们所着的服装制式与服色基本一致，只有头冠略有差异。

帻本是古时裹在头上的布，东汉时开始用一种平顶的帻作为戴冠时的衬垫，至西晋末年发展为前平后翘、只能罩住发髻的小冠，称为平巾帻（通常也称作小冠）。平巾帻在魏晋时造型较小，发展至唐则逐渐增大。壁画中官员头顶戴的平巾帻造型明显大于魏晋时期，已经接近盛唐的体量，其中左侧官员的平巾帻后端结构更加复杂，绘画整理时，基本遵照壁画上可辨的形态再现。按照当时的惯例，黑色平巾帻外罩漆纱笼冠、再插貂尾是一套标准而完整的搭配，但第335窟出现的这两位官吏中，右侧官吏未戴笼冠，只是头戴平巾帻簪貂尾，这一现象在湖北郧县李欣墓壁画中也曾经出现过。唐朝时期，头左侧簪貂尾的官员以散骑常侍为主，据此推断，壁画上的官吏可能是初唐的散骑常侍。

此外，左侧官员头戴的笼冠后侧还挑起一根条形饰物，细条自冠后经头顶垂至额前，最前端垂一穗状装饰，名为“垂笔”。沈从文先生在《中国古代服饰研究》中提及垂笔：本意是北朝时为了取法汉代细纱冠子与御史簪笔制度而制，因不得其法而出现了细条加垂穗的造型，并成为北朝文官服饰的“特别标志”。第335窟这位官吏的漆纱笼冠上插的垂笔，反映了初唐时官吏服饰体系中的北朝遗存现象。

（文：李迎军）

In Vimalakirti Sutra tableau on the north wall of the main chamber in cave 335, there are two ministers dating back to the Tang Dynasty who are listing with the princes of various countries. Although we can't distinguish their facial features clearly, according to their body language, they seem to be communicating with each other. Both of them wear square heart curved collar, Zhongdan and large sleeves robe. The style and color of the clothes they wear are basically the same, only the crown is slightly different.

Ze（帻）was originally a piece of cloth wrapped on the head in ancient times; since the Eastern Han Dynasty, it began to use a flat top as a cushion for wearing a crown. In the late Western Jin Dynasty, it developed into a small crown which was flat in front and upward in back and could only cover the bun. It was called Pingjinze（平巾帻, also called a small crown）. In the Wei and Jin Dynasties, the shape of Pingjinze was small, but the size gradually increased in the Tang Dynasty. In the mural, the shape of the Pingjinze on the top of the official's head is obviously larger than that of the Wei and Jin Dynasties, which is close to the size of the high Tang Dynasty. The structure of the back end of the Pingjinze on the left side of the mural is more complex. When doing the effect drawing, we basically followed the recognizable part of the mural, meanwhile according to the custom at that time, the black Pingjinze covered by Longguan（笼冠）and put a mink tail was a standard match. However, between the two officials in cave 335, the one on the right side does not wear Longguan, only wears Pingjinze with mink tail. This phenomenon also appeared in the painting of Li Xin's tomb in Yunxian County, Hubei Province. In the Tang Dynasty, the officials wearing mink tail on the left side of the head were mainly Sanqichangshi（散骑常侍, an official title, work beside emperor serves as counselor）. According to this, it can be inferred that the officials in the mural were probably Sanqichangshi in the early Tang Dynasty.

In addition, a strip ornament attached on the back of the Longguan wear by the official on the left side. The strip starting from the back of the crown reaches to the front of the forehead over the top of the head, and a tassel like decoration hanging at the front end, which was called "Chuibi"（垂笔, means hanging pen）. Mr. Shen Congwen mentioned Chuibi in *The Study of Ancient Chinese Clothes*: it originally appeared in the Northern Dynasty in order to follow the tradition of yarn crown and royal recorder hairpin pen of the Han Dynasty, but they did not fully understand, so the shape of thin strip and Chuibi appeared, which became the "special symbol" of civil official clothes in the Northern Dynasty. The Chuibi inserted in the lacquer yarn crown of this official in cave 335 reflected the Northern Dynasty tradition remains in the official dress system in the early Tang Dynasty.

（Text by: Li Yingjun）

图：李迎军 Picture by: Li Yingjun

The ministers' clothes on the north wall of the main chamber in cave 335 dated to the early Tang Dynasty at Mogao Grottoes

听法随侍大臣服饰 莫高窟初唐第335窟主室北壁

尽管在众多幅维摩诘经变中的各国王子听法图中，高句丽王子的位序始终偏后，以至于经常是只能露出上半身的形象，但高句丽王子也是听法图中出现得极为频繁的形象。《旧唐书·东夷》中记载了高句丽人的服饰情况："衣裳服饰……官之贵者，则青罗为冠，次以绯罗，插二鸟羽，及金银为饰。衫筒袖，裤大口，白韦带，黄韦履。"书中的描述可以与壁画中的人物形象对照比较，因其具备颇有特点的服饰表现，所以在相关文献与图像的佐证下，可以基本明确他们的民族属性。

第335窟主室北壁的高句丽王子为二人，皆头戴插双鹖尾小冠，以鸟羽装饰头冠是高句丽服饰的典型特征，鹖性凶猛，以鹖尾饰冠有彰显勇猛之意。二人皆穿着大袖袍服，领口、袖口有宽缘饰，只是服色有异。听法图中的二位使臣位列众人之中，下半身的形象由于被其他王子遮挡而没能在壁画上体现。参照职贡图、礼宾图的资料，可知高句丽男装与小冠、大袖袍服搭配穿着的应是下穿大口裤、腰束革带、足蹬皮靴。

（文：李迎军）

Because the Koguryo prince always stands backward among other princes in Vimalakirti Sutra tableau, so that they often only show their upper body, Koguryo prince is a popular figure in this kind of story. In the book *the old Tang Dynasty-the east area*（《旧唐书·东夷》）: the records of the Koguryo people: "the Koguryo people's clothes...The high rank officers wear Qingluo, the lower rank wear Feiluo, also decorated with two bird feathers, gold and silver. The shirt sleeves are like tubes, the trousers are wide at the ends, white belt and yellow shoes." The description in the book can be compared with the figures in the murals. Because of their distinctive clothes appearance, their national attribution can be basically clarified with the support of relevant literature and images.

There are two Korean princes on the north wall of the main chamber in cave 335. They all wear small crowns with double crossoptilon mantchuricum feathers. The crowns decorated with bird feathers are typical characteristics of Korean clothes. Crossoptilon mantchuricum is ferocious in nature, and the crowns are decorated with the tail meaning bravery. Both of them wear large sleeves robe with wide trimming at the neckline and cuffs, except the colors are different. The two envoys listening to Dharma in the picture of are standing among people, so the lower part of their bodies are not reflected in the mural because it is blocked by other princes. Referring to the information in the *Official Tribute Picture*（《职贡图》）and *the Meeting Guests Picture*（《礼宾图》）, we know that the matching of Koryo man's clothes with small crown and large sleeves robe should be big mouth pants, leather belt around the waist and leather boots.

（Text by: Li Yingjun）

图：李迎军 Picture by: Li Yingjun

帝王听法图蔽膝图案

莫高窟初唐第335窟主室北壁

The pattern on the emperor's Bixi

on the north wall of the main chamber in cave 335 dated to the early Tang Dynasty at Mogao Grottoes

此幅蔽膝图案来源于第335窟主室北壁的帝王听法图中的帝王冕服。作为冕服的组成部分，蔽膝又有“韨”“韍”“芾”的别称。《左传》桓公二年，郑玄注曰：“芾，太古蔽膝之象，冕服谓之芾。”又有诗云：“朱芾斯皇”“赤芾在股”，则芾是当股之衣，穿法是系与革带之上而垂直于膝盖前，因而得名“蔽膝”。

图案结构为网状连缀形，是由单列的对波纹组合而成。对波形式是丝绸纹样中一种骨架构图的名称，盛行于南北朝至隋唐时期，同时见于建筑装饰与服饰面料中。由单列对波组合成的连缀式四方连续纹样，表现在丝绸提花纹样上主要有两种形式：一种是折枝纹样的连缀，另一种是几何纹样的连缀。蔽膝主体的波状曲线正反交错形成近菱形的连缀图案，这种图案流行于唐代织物上。这种装饰骨架与建筑装饰、壁画、织物纹样上都有着同一性的传承关系。

织物上方条带表示革带，有两列菱格图案。蔽膝图案由石绿色三向包围，主体纹样为几何菱形网状连缀纹，于网状中嵌以十字形三瓣四叶的小花，这种十字形嵌花，与唐代印染织物中常见的放射式的团花、方胜纹、小簇草花一样，体现在敦煌壁画的服饰中。

（文：高雪）

This pattern comes from the emperor's ritual clothes in the picture of emperor listening to Dharma on the north wall of the main chamber in cave 335. As an part of the ritual dress. Bixi is also called “韨(Fu)”, “韍(Fu)” and “芾(Fu)”. In the second year of Duke Huan（桓公）recorded in *Zuozhuan*（《左传》）, Zheng Xuan wrote: “Fu is the covering for knees in the ancient times, in emperor's ritual clothes is called Fu.” Another poem says: “Red Fu is for emperors” and “Red Fu covers the legs”. So Fu is the part which covers the legs' front side, and the wearing method is tied by leather belt and drape to knees, so it is named “Bixi, which means covering knees”.

The general structure of the pattern is a net, which is composed by single row of corrugations. Opposite corrugation form（对波形式）is the name of a kind of frame composition in silk patterns, which was popular in the Southern and Northern Dynasties to the Sui and Tang Dynasties, and also found in architectural decoration and clothes fabrics. it is four in a group repeated pattern composed by single row opposite corrugation. In silk jacquard patterns, there are two main forms: one is the combination of folded branch patterns, the other is the combination of geometric patterns. The wavy curves on Bixi are crisscrossed to form a nearly rhombic pattern, which is popular on the fabrics of the Tang Dynasty. This kind of decorative frame has the same inheritance relationship with architectural decoration, murals and fabric patterns.

The upper stripes of the fabric represent the leather belt with two rows of rhombic patterns. The Bixi pattern is surrounded by malachite green in three directions. The main pattern is a geometric rhombic reticular pattern. The reticular pattern is inlaid with small cross shaped flowers with three petals and four leaves. This kind of cross shaped inlaid flower is the same as the radial flower clusters, Fangsheng patterns（方胜纹）and small clusters of grass flowers are commonly seen in the printing and dyeing fabrics in the Tang Dynasty. They are reflected on clothes in Dunhuang murals.

（Text by: Gao Xue）

图：高雪　Picture by: Gao Xue

敦煌莫高窟初唐

第372窟

Dunhuang Mogao Grottoes Cave 372 of the Early Tang Dynasty

第372窟是初唐时期建造的覆斗顶小型窟。窟顶藻井为团花井心，边饰为卷草纹，有垂幔，四披绘千佛。主室西壁开一平顶敞口龛，龛内清塑一铺五身。龛内浮塑背光两侧绘二赴会佛、二菩萨、二弟子、二飞天，龛顶绘宝盖和二化佛。龛外两侧各绘一菩萨。主室北壁绘说法图一铺，南壁绘阿弥陀经变。东壁门南北分别为地藏王和药师佛。前室和甬道存宋代壁画。

西壁龛外两侧存留菩萨两身，服饰层次较分明。南壁阿弥陀经变中有多身菩萨服饰清晰，结构十分复杂，服饰纹样非常精美，其中有多种菱格纹、小团花、扎经染等，石绿色在整体色调中突出并稳定。

（文：崔岩）

Cave 372 is a small truncated pyramidal ceiling cave built in the early Tang Dynasty. The caisson on the top of the cave has a round flower pattern as the center. The edges of the caisson are decorated with scrolling grass pattern and fringes, and the four slopes are covered by Thousand Buddha motif. In the west wall of the main chamber, there is an open niche with a flat top, in which there are five painted sculptures made in the Qing Dynasty. The back light is relief, on the two sides are painted two attending Buddhas, two Bodhisattvas, two disciples and two flying Apsarases. The top of the niche is painted with a canopy and two Buddhas, both outer sides of the niche have a painted Bodhisattva. The north wall of the main chamber is painted a picture of preaching scene, while the south wall is painted Amitabha Sutra tableau. The north and south side of the east gate are Ksitigarbha and Bhaisajyaguru. There are murals dating back to the Song Dynasty in the front room and corridor.

There are two painted Bodhisattvas on both outer sides of the west niche, and their clothes are well-defined. In the Amitabha Sutra tableau on the south wall has many Bodhisattvas and their clothes are clear to see, complex in structure, and exquisite in patterns, such as rhombic patterns, small round flowers, tie-dyeing and so on. The malachite green is prominent and stable in the overall tone.

(Text by: Cui Yan)

The pattern on Manjusri Bodhisattva's antariya

on the south wall of the main chamber in cave 372 dated to the early Tang Dynasty at Mogao Grottoes

文殊菩萨裙身图案

莫高窟初唐第372窟主室南壁

这里的图案取自主室南壁阿弥陀经变中文殊菩萨的裙饰。裙子材质为透明纱质，因此绘制整理时以肤色为底色，以十字散花纹为显花。主纹为十字结构，浅石绿色花瓣，白线勾勒，呈四方连续排列。纹样清新雅致，花地分明，清地排列。此类十字散花纹在敦煌石窟中多次出现，是一种基础性服饰图案。

（文：崔岩、苏芮）

The pattern here is taken from the Manjusri Bodhisattva's antariya in Amitabha sutra tableau on the south wall of the main chamber. The antariya is made of transparent silk, so the skin color is used as the background color and the cross pattern is used as the main pattern when drawing effect picture. The main pattern is a cross structure with light malachite green petals, outlined by white lines, arranged in four as a group repeated. The pattern is fresh and elegant, the flower shape is clear, and the arrangement is simple. This kind of cross pattern appeared many times in Dunhuang Grottoes, which was a basic dress pattern.

(Text by: Cui Yan, Su Rui)

图：苏芮 Picture by: Su Rui

1—2

敦煌莫高窟初唐

第375窟

Dunhuang Mogao Grottoes Cave 375 of the Early Tang Dynasty

第375窟是初唐时期开凿的覆斗顶窟。主室窟顶藻井绘石榴莲花井心，有伎乐飞天与垂幔铺于四披，四披绘千佛。主室西壁开平顶方口龛，龛内彩塑一佛、二弟子、四菩萨，均经清代重修。龛壁中央绘火焰纹背光，下绘婆薮仙和鹿头梵志，南北侧各绘三身菩萨，龛顶绘飞天九身，上身半裸，下着长裙。龛外南北两侧上部分别绘夜半逾城和乘象入胎等佛传故事，中绘菩萨各一身，下部为供养比丘、比丘尼、男女供养人及侍从。龛下为五代绘供器和供养菩萨。主室南北壁绘千佛，中央各绘说法图一铺，下绘女供养人十一身、侍从二十身及男供养人十四身、侍从二十身。东壁门上绘千佛，门南北各绘天王二身，下为五代绘供养比丘尼、女供养人及侍从。前室和甬道现存五代壁画。西壁龛外南北两侧菩萨与南北两壁说法图中菩萨服装造型结构依赖长线表现，整体修长。

此窟主室南北壁下部保留的多身男女供养人及侍从壁画对于研究初唐时期服饰文化尤为重要。女供养人像造型较为瘦削，所着服饰突出了女性体态身姿的修长窈窕，虽与陕西乾县永泰公主墓壁画中侍女的服饰相类似，但显得更加紧窄和保守。男供养人像体现出当时男子着幞头、圆领袍、革带、乌靴等服饰搭配的流行趋势，从软脚幞头、袍服长度等处可以发现从隋代向唐代逐渐发展的服饰文化进程。主室西壁龛外壁画以土红为底色的佛传故事画，在构图、色彩和造型上保留了隋代遗风，这在初唐洞窟中不太多见，反映了艺术风格和审美习惯在历史转变时期的风貌。

（文：崔岩）

Cave 375 is the truncated pyramidal ceiling cave excavated in the early Tang Dynasty. The caisson on the top of the main chamber is painted with pomegranate and lotus in the center. There are musician flying Apsarases and fringes reach to the four slopes, and the four slopes are painted with Thousand Buddha motif. The west wall of the main chamber has a square opening niche with a flat top. In the niche, there are painted sculptures of one Buddha, two disciples and four Bodhisattvas, all of which were repainted in the Qing Dynasty. On the central part of the niche wall is painted with flame pattern backlight; the lower parts are painted with the Images of Vasistha（婆薮仙）and Mrgasirsa（鹿头梵志）; both north and south sides are painted with three Bodhisattvas; the top of the niche is painted with nine flying Apsarases; their upper bodies are naked, and the lower parts are wearing long antariya. On the north and south outside of the niche, the upper parts are painted Buddhist stories such as Great Departure and Great Conception. The middle parts are painted with a Bodhisattva image on each side, and the lower parts are painted with monks, nuns, male and female donors and attendants. Below the niche is painted with offering vessels and offering Bodhisattvas dating back to the Five Dynasty. The north and south walls of the main chamber are painted with Thousand Buddha motif, and each side has a picture of preaching scene in the center. The lower part is painted with 11 female donors, 20 attendants and 14 male donors, 20 attendants. There is Thousand Buddha motif painted above the east gate, two heavenly kings painted both on the north and south side of the gate, and the lower part has nuns, female donors and attendants images dating back to the Five Dynasties. There are murals dated to five dynasties in the front room and corridor. The Bodhisattvas on the north and south outside of the west niche and on the north and south side of the main wall, their clothes structure in the picture depends on the long-term lines and their figures are slender in general.

The murals on the lower part of the north and south walls of the main chamber in this cave are particularly important for the clothes culture study of the early Tang Dynasty. The female donors figures are thin in shape, and the dress highlights the slender and graceful body. Although it is similar to the maid dress in the painting on the Princess Yongtai（永泰）tomb wall at Qianxian（乾县）County, Shanxi Province, this style is more narrow and conservative. The male donors reflected the fashion trend of man wearing Fu headscarf, round collar robe, leather belt and black boots at that time. From the soft Fu headscarf and robe length, we can see the changing period of clothes culture from the Sui Dynasty to the Tang Dynasty. The murals outside the west niche of the main chamber used earth red as the background, and painted Buddhist stories; the color and shape retains the Sui Dynasty tradition in composition, which is rare in the early Tang Dynasty caves, reflecting the art and aesthetic habits change in the historical transition period.

（Text by: Cui Yan）

女供养人服饰

莫高窟初唐第375窟主室南壁下

The female donor's clothes on the lower south wall of the main chamber in cave 375 dated to the early Tang Dynasty at Mogao Grottoes

第375窟南壁所绘女供养人与侍从的服饰虽有色彩等差异，但造型风格统一，从中可窥初唐女装风貌。在服饰效果图整理时，选取了保存相对完整的一组形象，为一女供养人偕两女随侍。女供养人手执香炉，头梳高髻，服饰为初唐典型风格，上身穿石绿色窄袖交领上襦，披帛自双肩绕臂自然垂于两侧；束腰带，下着石青色高腰长裙，裙长及地，着绣花履。此窟壁画中的女子形象多衣裙窄小，更显身材颀长，唐初女装流行窄袖小衫，与北周、隋代相近，只是衣袖长短不同而更趋合身。供养人身后跟随两侍从，一侍从上身着浅土黄交领窄袖小衫，披披帛，下着拖地长裙。另一侍从上着圆领小袖齐膝长袄子，下着条纹小口裤，脚穿软底锦鞦靴，一般来说，这种穿着明显受西域或波斯影响，多为宫女或身份较低女侍，实即“女扮男装”，若上身搭配翻领小袖长袄子，即为胡服新装演绎式样。

初唐稍后，身份较高的贵族妇女发髻已一改隋代的平云式单纯，向上高耸，发展成种种不同变式。《妆台记》载：“唐武德中，宫中梳半翻髻，又梳反绾髻、乐游髻。”新奇样式上行下效，一时成为风气。有大臣曾请唐太宗下令禁止，唐太宗初加以斥责，但后来询及近臣令狐德棻，问妇女发髻加高是什么原因？令狐德棻以为，头在上部，地位重要，高大些也有理由，因此高髻不受任何法令限制，逐渐更加多样化。壁画中女供养人的发髻有向上腾举之势，被文人形容“离鸾惊鹄之髻”，或即“回鹘髻”，是唐初流行式样中比较主要的一种发髻。

（文：吴波）

Although the clothes of the female donors and attendants on the south wall of cave 375 are different in color, the styles are generally same, from which we can see the style of the early Tang Dynasty woman's dress. In the process of drawing the effect picture, we choose this group in which the images are relatively complete, a female donor and two female attendants. The female donor is holding a incense burner, combed the hair a high bun, dressed typical early Tang Dynasty clothes. She wears a malachite green narrow sleeves dress, a uttariya cover from her shoulders to arms, naturally hanging on both sides. She wears a belt, a azurite high waist antariya, and embroidered shoes. In the murals of this cave, women's clothes mostly are narrow and small, which made the figures look more slender. In the early Tang Dynasty, women's clothes were popular with narrow sleeves and small dress, which were similar to those of the Northern Zhou Dynasty and the Sui Dynasty, but the length of sleeves were different and more suitable. Two attendants stand behind the donor; one of them is wearing a light yellow cross collar shirt with narrow sleeves, a uttariya and a floor dragging antariya; the other one is wearing a knee length jacket with round neck and narrow sleeves, striped trousers and soft bottom brocade boots. Generally speaking, this kind of dress was obviously influenced by the Western Regions or Persia. Most of them were palace maids or maids with low status. In fact, they were "women disguised as men". If the upper clothes changed to lapels and narrow sleeves long jacket, it was a new style of Hu dress.

Later in the early Tang Dynasty, the noble woman's hair bun had changed from the Sui Dynasty's plain cloud style to a variety of different styles. The story of makeup（妆台记）states: "In year Wude of the Tang Dynasty, the palace woman combed the hair in Banfan style（半翻髻）, Fanwan style（反绾髻）, and Leyou style（乐游髻）." These new styles spread widely and then became popular for a while. A minister once asked Emperor Taizong to order a ban, and at the beginning, Emperor Taizong reprimanded this trend, but later he asked his close minister Linghu defen（令狐德棻）why woman's hair was raised? Linghu defen responded that the head is on the upper part of body and important, using hair to make it taller and bigger is understandable and acceptable. Therefore, high hair bun was not restricted by any laws and regulations, and gradually became more diversified. In the mural, the bun of the female donor has the tendency of rising, which was described by the literati as "the bun like phoenix and swan on the head", or Uighur bun, which was the main popular style in the early Tang Dynasty.

（Text by: Wu Bo）

图：吴波 Picture by: Wu Bo

The male donor’s clothes

on the lower north wall of the main chamber in cave 375 dated to the early Tang Dynasty at Mogao Grottoes

男供养人服饰

莫高窟初唐第375窟主室北壁下

敦煌莫高窟唐代壁画中的世俗人物，尤其是供养人的服色明显比北朝和隋代的丰富得多。第375窟男供养人与侍从的服装为一类，都是头裹软脚幞头，穿圆领袍服，腰束革带，足蹬皮靴。每人面容神采各异：主人多雍容华贵，手捧盘瓶、香炉、鲜花，或双手拱于胸前，面容恭谨虔诚。侍从表情更为丰富，有的手捧贡品紧随主人，有的身体前倾翘脚欲行，有的手扶斜背的包裹回头顾盼，有的戴风帽双手拱于胸前四处张望，与主人的肃穆相比，神情、动态活泼，富有年龄特点。此组供养人与侍童组画，较好地还原出当时场景。

（文：吴波）

In the Tang Dynasty murals of Mogao Grottoes in Dunhuang, the clothes color of secular figures, especially the donors, are much richer than those of the Northern Dynasties and the Sui Dynasty. The clothes of the male donors and attendants in cave 375 are of the same type. They all wear soft feet Fu hat, round collar robes, leather belt around the waist, and leather boots. Each person’s facial expression is different: the donors are elegant, holding plate and bottle, censer, flowers, or hands in front of their chest, the expressions are respectful and devout. The attendants have more colorful expressions; some hold tribute close to the master, some lean forward to walk, some grab package and look back, and some wear wind hat and put their hands in front of chest and look around. Compared with the solemnity of the masters, they have more dynamic and lively look, full of age characteristics. This group of donors and the waiters images are very realistic for the scene at that time.

(Text by: Wu Bo)

图：吴波 Picture by: Wu Bo

The male donor’s clothes
on the lower north wall of the main chamber in cave 375 dated to the early Tang Dynasty at Mogao Grottoes

男供养人服饰
莫高窟初唐第375窟主室北壁下

幞头是受鲜卑帽影响，同时也是少数民族帽式与汉族幅巾相互交融之后的产物。渊源于北魏，初创于北周，成型于隋，盛行于唐，历宋、元、明，直到清初被满式冠帽取代，幞头及其变体，通行了整整一千余年，是此时期男装的独特标志。在隋唐时期，幞头脚经历了由短而长、由窄而宽、由软而硬的变化。隋到初唐的幞头皆为两脚短而软，后来将其加长，所谓“长脚罗幞头”仍然为柔软材质。

唐代圆领袍服是当时最具代表性、最为流行的男装，它是在旧式鲜卑褶衣的基础上，吸收西域胡服和中原袍服融合而成，产生于北朝晚期的一种服装。

（文：吴波）

Fu hat was influenced by Xianbei hat, and it was also the result of the combination of Han hat and minority hat. It originated in the Northern Wei Dynasty, appeared in the Northern Zhou Dynasty, shaped in the Sui Dynasty, popular in the Tang Dynasty, then pass to the Song Dynasty, Yuan Dynasty, Ming Dynasty; until the early Qing Dynasty, it was replaced by the Man style cap. Fu hat and its variants had been popular for more than 1000 years, which is a unique symbol of man’s clothes in this period. In the Sui and Tang Dynasty, Fu hat feet experienced changes from short to long, from narrow to wide, from soft to hard. From the Sui Dynasty to the early Tang Dynasty, Fu hat had short and soft feet. Later, it was lengthened. The so-called “long feet Luofutou（长脚罗幞头）” was still made of soft material.

The round collar robe in the Tang Dynasty was the most representative and popular man’s clothes at that time. It was based on the old Xianbei Zhe clothes, absorbed the Hu clothes of the Western Region and combined with the Central Plains robe clothes, appeared in the late Northern Dynasty.

(Text by: Wu Bo)

图：吴波 Picture by: Wu Bo

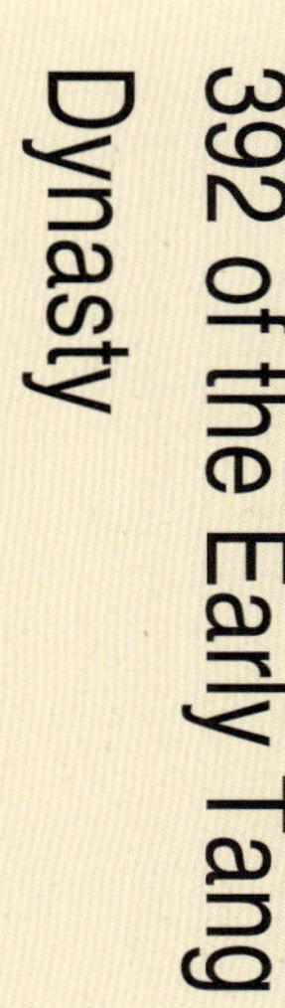

Dunhuang Mogao Grottoes Cave 392 of the Early Tang Dynasty

敦煌莫高窟初唐

第392窟

第392窟是隋代创建的覆斗顶洞窟，经初唐、中唐、五代、清代重修。根据《敦煌石窟内容总录》的研究，此窟内被确定为初唐时期壁画的是主室西壁双层方口龛内外层龛壁两侧所绘弟子像，隋代壁画仅保留在主室窟顶、西壁和东壁门上，隋代彩塑也经清代重修，其他壁画均为中唐或五代时期所作。因为此窟开凿和修建时间较长，跨越不同时代，所以壁画和彩塑的布置规划缺乏完整性，显得略微散乱。

主室西壁龛内有隋代彩塑一佛、二弟子、四菩萨，塑像头部、手部及部分服饰经清代重修，已失去原貌。但是通过对服饰图案题材、构图、色彩和风格的分析，可以发现部分彩塑服饰仍保留了初唐时期重绘的面貌，具有一定的典型性，因此选取此部分图案进行整理绘制。西龛主尊佛服饰图案和双层龛外层南侧菩萨服饰图案，与西龛内层龛边饰图案造型风格统一，主体为疏朗的初唐卷草，加十字结构团花和半团花，色彩以石绿和白色为主，十分清新典雅。

（文：崔岩）

Cave 392 is truncated pyramidal ceiling cave built in the Sui Dynasty and rebuilt in the early Tang dynasty, middle Tang dynasty, Five Dynasties and Qing Dynasty. According to the research of *the Dunhuang Grottoes Content General Record* (《敦煌石窟内容总录》), the mural in this grotto dating back to the early Tang Dynasty are the portraits of disciples painted on both sides of the inner and outer niche of the double-layer square opening niche on the west wall of the main chamber. The Sui Dynasty mural are only kept on the top of the main chamber, the west wall and above the east door. The painted sculptures of the Sui Dynasty were also repainted in the Qing Dynasty. Other murals were painted in the middle Tang Dynasty or the Five Dynasties. Because the excavation and construction of this cave took a long time and spanned different times, the layout of murals and painted sculptures lacked integrity and appeared a little disorder.

In the west niche of the main chamber, there are Sui Dynasty painted sculptures of one Buddha, two disciples and four Bodhisattvas. The head, hands and some clothes of the sculptures have lost their original appearance after being repainted in the Qing Dynasty. However, through the analysis of the theme, composition, color and style of dress patterns, it can be found that some of the painted clothes still retain the appearance of the redrawn in the early Tang Dynasty, which has a certain typicality. Therefore, this part of the pattern is selected for drawing and studying. The clothes pattern of the main Buddha and the Bodhisattva on the south side of the outer layer of the double-layer niche are unified with the patterns on the inner niche edges. The main body is the scrolling grass pattern of the early Tang Dynasty, plus the cross structure of flower medallion and half flower medallion. The colors are mainly malachite green and white, which is very fresh and elegant.

(Text by: Cui Yan)

The pattern on painted sculpture Buddha's kasaya in the west niche of cave 392 dated to the early Tang Dynasty at Mogao Grottoes

彩塑佛陀袈裟图案 莫高窟初唐第392窟西壁龛内

这身彩塑佛像的头部、手部和皮肤经清代重绘，所幸部分袈裟图案还保留了初唐时期原妆，其服饰图案十分丰富，包括卷草纹、宝相花纹、缠枝纹等。

外层袈裟的缘边和主体为卷草纹，这是敦煌莫高窟代表性的植物纹样之一，由早期的忍冬纹发展变化而来，因兴盛于唐代，又称“唐卷草”。其特点为枝干呈卷曲波浪状，有疏密两种分布，呈二方连续或四方连续等不同形式，由于结构的灵活性，能够适应藻井边饰、人物服饰等更多的装饰空间，在同时期的金银器、碑刻、织锦、瓷器上也有表现。关于卷草纹的源流有不同说法，但对卷草纹的地位和价值鲜有异议。卷草纹的造型是以波浪式结构线作为骨干，花头呈等距、错落排列，花叶通过波浪形枝干连接和分布，疏畅有度，形式曲折优美，生动灵巧。敦煌莫高窟古代画工讲究曲率和笔意，熟练掌握自然界的花草特征，对纹饰的张力造型熟练于心，在纹饰中表现出了强大的生命力。

内层袈裟缘边为两种不同底色和造型的宝相花纹。宝相花纹是丝绸之路上本土艺术形式与外来艺术不断融合过程中，形成的具有民族审美特色的植物花卉，取自佛教“庄严宝相”之名，据文献记载“宝相”一词可追溯至魏晋南北朝时期。宝相花纹样受古印度犍陀罗艺术的熏染，又与中国传统装饰艺术与地域审美相结合，寓意富贵吉祥，符合民族审美和风俗。从图案造型上来看，宝相花纹概括了不同花体（牡丹、石榴花等）的特征进行艺术再加工，使纹样形状更趋于丰富化、平面化，整体造型圆曲丰满。绘制时以“米”字为基调，呈平面圆形辐射状，土红底色的宝相花纹与陕西乾县唐代懿德太子墓壁画中方形的宝相花纹类似，均以圆心为顶点，内圈的花瓣组合成圆环状，外圈则明显呈方形；白底色的宝相花纹的花心画四片如意纹，外饰八片花瓣，第三、第四层以八片如意、花头为主，如意纹与花瓣交错，层层叠压形成有节奏的轮廓形态。

（文：姚志薇）

The head, hands and skin of this painted Buddha sculpture were repainted in the Qing Dynasty. Fortunately, some kasaya patterns still retain the original appearance of the early Tang Dynasty. The decoration patterns on this kasaya are very rich, including scrolling grass pattern, Baoxiang pattern, rolled-vine pattern and so on.

The trimming and main body of the outer kasaya are scrolling grass pattern, which was one of the representative plant patterns in Dunhuang Mogao Grottoes. It was derived from the early honeysuckle pattern and is also called "Tang scrolling grass（唐卷草）" because it flourished in the Tang Dynasty. Due to the flexibility of the structure, it can be adapted to more decorative space, such as caisson ornaments, clothes and so on. It also appeared on gold and silver ware, tablet, brocade and porcelain of the same period. There are different opinions about the origin of scrolling grass pattern, but there are few disputes about the status and value of the this pattern. The shape of the scrolling grass pattern is based on the wavy structure line, two kinds of density distribution, dense or sparse two in a group continuous or four in a group continuous forms. The flower heads are equidistant and arranged oppositely, and the flower leaves are connected and distributed through the wavy branches, which is sparse and smooth, tortuous and beautiful, vivid and dexterous. The ancient painters at Dunhuang Mogao Grottoes paid attention to the curvature and brushwork, mastered the characteristics of flowers and plants in nature, and were proficient in the modeling of patterns, showing a strong vitality in pattern creation.

The trimming of the inner kasaya has two kinds of Baoxiang patterns with different background colors and shapes. Baoxiang pattern is a flower pattern with national aesthetic characteristics formed in the process of the repeated integration of local art forms and foreign art on the silk road. It comes from the name of "solemn and precious form（庄严宝相）" in Buddhism. According to the literature, the word "Baoxiang" can be traced back to the Wei, Jin, Southern and Northern Dynasties. Influenced by Gandhara art of ancient India, the pattern of Baoxiang flower is combined with traditional Chinese decorative art and regional aesthetics, implying wealth and auspiciousness, which is in line with national aesthetics and customs. From the point of view of pattern modeling, Baoxiang pattern integrated the characteristics of different flower bodies（peony, pomegranate flower, etc.）for artistic reprocessing, so that the pattern shape tends to be more rich and flat, and the overall shape is round and full. The Baoxiang pattern is based on the "米" shape, performed as radial circular. The red earth Boxiang flower pattern is similar to the square pattern on the tomb wall of Prince Yide of Tang Dynasty in Qianxian County, Shaanxi Province, with the center of the circle as the vertex, the petals on the inner ring are combined into a circular ring, and the outer ring is obviously square; the pattern on the white background has four Ruyi patterns in the center, eight petals on the outside, and eight petals on the third and fourth layers. The Ruyi pattern is interlaced with petals to form a rhythmic outline.

（Text by: Yao Zhiwei）

图：姚志薇 Picture by: Yao Zhiwei

The pattern on the painted Bodhisattva's antariya

on the south side of the outer niche on the west wall of the main chamber in cave 392 dated to the early Tang Dynasty at Mogao Grottoes

彩塑菩萨裙身图案

莫高窟初唐第392窟主室西壁外层龛南侧

这是第392窟主室西壁外层龛南侧的彩塑菩萨，始塑于隋代，上身基本都由清代重修，改变了原本的面貌，所幸裙身图案仍然保留了初唐风格。

图案的整体构图采用初唐菩萨裙饰的常用方式，即从双股处延伸两道竖向图案，裙子下部装饰横向的多道边缘图案，与初唐第334窟主室西壁彩塑菩萨的裙身图案构图类似，反映了初唐时期服饰图案的定位设计思想。图案均为二方连续式结构，题材为唐代流行的卷草纹和团花纹，单位分布排列较为疏朗。图案色彩淡雅明快，在初唐时期的服饰图案中是较为特殊的一例。

（文：崔岩）

This is the painted sculpture Bodhisattva on the south side of the outer niche on the west wall of the main chamber in cave 392. It was first molded in the Sui Dynasty. The upper body was basically repainted in the Qing Dynasty, which has already changed its original appearance. Fortunately, the antariya design still retains the style of the early Tang Dynasty.

The overall composition of the pattern adopts the common way of Bodhisattva antariya decoration in the early Tang Dynasty; that is, two vertical patterns are extended from the waist, and the lower part of the antariya is decorated with horizontal multiple trimming patterns, which is similar to the pattern composition of the Bodhisattva antariya in cave 334 on the west wall of the main chamber dating back to the early Tang Dynasty, reflecting the positioning design idea of dress in the early Tang Dynasty. The motif is the popular scrolling grass pattern and flower medallion pattern in the Tang Dynasty, and the arrangement of the units are sparse. The color of the pattern is elegant and bright, which is a special example in the early Tang Dynasty.

(Text by: Cui Yan)

图：崔岩　Picture by: Cui Yan

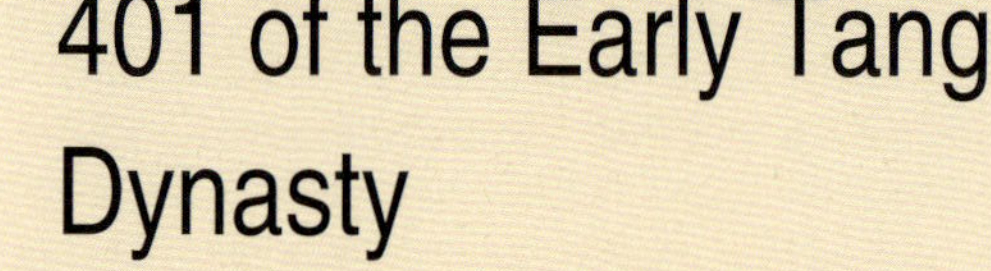

敦煌莫高窟初唐
第401窟

Dunhuang Mogao Grottoes Cave 401 of the Early Tang Dynasty

第401窟是创建于隋代的覆斗顶窟，经初唐、五代、清代重修。初唐壁画主要包括：主室西壁所开双层方口龛外层南北壁下部各绘二身供养菩萨，龛外下部绘十身供养菩萨；主室南壁西侧方口斜顶圆券龛内壁绘背光、六菩萨、二飞天、二赴会佛、一持花供养童子，龛下绘阿弥陀佛一铺、男女供养人各二身，东侧存供养菩萨三身；主室北壁西侧方口斜顶圆券龛下说法图一铺及两侧大力菩萨、满月菩萨等五身供养菩萨；东壁门上绘说法图一铺，门南北两侧中部各绘说法图一铺，下为辩音菩萨、虚空藏菩萨、常举手菩萨、常下手菩萨等供养菩萨像。

可见，初唐时期主要对此窟壁面下部进行重点重绘，特别是绕窟一周添绘了众多供养菩萨像。虽然有些颜料已经变色，但是可以看出这些供养菩萨面相丰腴，神情虔诚，姿态各异，身形婀娜，服饰飘逸。她们或持莲花，或托净器，在漫天花雨中信步踏于莲花座上。此窟中菩萨的服饰由于肩头垂下的飘带、披帛、璎珞与长裙细密的褶皱等方面原因，整体造型线结构特征突出，修长舒展，并重点描绘了面料的装饰纹样。

（文：崔岩）

Cave 401 is a truncated pyramidal ceiling cave built in the Sui Dynasty and repainted in the early Tang dynasty, Five Dynasties and Qing Dynasty. The mural dating back to the early Tang Dynasty mainly include: the double-layer square mouth niche in the west wall of the main chamber, two painted offering Bodhisattvas on both outer side of the niche, lower parts of the north and south walls, and ten painted offering Bodhisattvas below the niche; the inner wall of the square mouth oblique top round edge niche in the west side of the south wall in the main chamber is painted with backlight, six Bodhisattvas, two flying Apsarases, two attending Buddhas and one offering child holding flowers; below the niche is painted with Amitabha Buddha, two male and two female donors, and three painted offering Bodhisattvas on the east side. On the west side of the north wall of the main chamber, below the square mouth oblique top niche has a preaching scene, beside the niche have five offering Bodhisattvas such as the Great Power Bodhisattva, the Full Moon Bodhisattva. Above the east door, there is a picture of preaching scene, and on the north and south side of the door, there is also a picture of preaching scene each in the middle area. On the lower areas have offering Bodhisattvas, such as the Voice Distinguishing Bodhisattva, the Void Bodhisattva, the Hand Raising Bodhisattva and the Hand Lowering Bodhisattva.

It can be seen that in the early Tang Dynasty, the lower part of the wall in this cave was mainly redrawn, especially a large number of offering Bodhisattvas were painted around the cave. Although some of the colors have changed, we still can see that these Bodhisattvas are plump, devout, graceful and elegant. They either hold a lotus, or hold a water vase, and walk on the lotus seat in the flower rain. The Bodhisattva's clothes in this cave are characterized by the ribbon hanging from the shoulder, uttariya, keyūra and long antariya's fine folds, etc. The structure of the overall modeling line is prominent, slender and stretch, and the decorative patterns of the fabric are mainly depicted.

(Text by: Cui Yan)

The offering Bodhisattva's clothes

on the lower area of the north wall in the main chamber of cave 401 dated to the early Tang Dynasty at Mogao Grottoes

供养菩萨服饰

莫高窟初唐第401窟主室北壁下

初唐时期，在第401窟主室四壁下部重绘了多身供养菩萨像，均向着主尊的方向行进，但具体姿态各不相同，这里绘制整理的是主室北壁下部东端的一身供养菩萨。菩萨为正面像，立于双层莲花座上。她面容端庄秀美，双目微合，向下凝视，体态轻盈，脉脉含情。左手轻提薄纱天衣一端，右手托玻璃宝珠盘，盘子晶莹剔透，呈浅蓝色，口沿有八个褐色圆钮装饰，应是当时十分珍贵的工艺品。菩萨身体重心略放于右腿，增加了飘然而至的态势。

服饰造型上，菩萨头戴嵌珠宝冠，戴璎珞和手镯，璎珞在肩部镶嵌宝珠并垂下长短不一的飘带。上身斜披土红色披帛，上面装饰菱格联珠纹图案，青色透明的天衣随身，腰中束带，裙腰处翻折出波状衣褶。长裙落地，在两侧由璎珞珠串兜起圆弧，令裙子的整体造型增加了起伏变化。璎珞和天衣随风而动，搭配身边飘落的散花，整体上给人以天女下凡的视觉美感。在画面艺术处理上，线条流畅洒脱，刚柔并济；色彩配置以赭褐暖调为主，青绿穿插点缀其间。

（文：刘元风）

In the early Tang Dynasty, the lower area of the four walls in the main chamber of cave 401 was redrawn with multiple offering Bodhisattvas, all of which seem moving towards the main Buddha, but the individual postures are different. Here is the offering Bodhisattva at the east end on the lower area of the north wall in the main chamber. This Bodhisattva is standing on the double lotus seat and facing to us; her face is dignified and beautiful, and her eyes are slightly closed; she gazes down, her body is light and affectionate. One end of the thin silk clothes is gently lifted by the left hand, and a glass pearl plate is held by the right hand. The plate is crystal clear, light blue, and decorated with eight brown round buttons along the edge. It should be a very precious handicraft at that time. The weight of the Bodhisattva's body is slightly placed on her right leg, which increases her flying momentum.

In terms of dressing style, the Bodhisattva's head is decorated with a jewel crown, wearing keyūra and a bracelets. The keyūra is decorated with jewels on the shoulder and has streamers of different lengths. The upper body is obliquely covered with earth red uttariya, which is decorated with diamond lattice beads pattern. The blue transparent heavenly clothes suit the body, hold by a belt on the waist, and the antariya waist wrap is folded into wavy pleats. The long antariya falls to the ground, and the two sides of the antariya hold up by keyūra in circle shape, which increases the fluctuation of the overall shape of the antariya. Keyūra and heavenly clothes move with the wind, together with the scattered flowers around her, making people feel the visual beauty of celestial woman descending to earth. In the artistic treatment of the picture, the lines are smooth and free, with the combination of hardness and softness; the color configuration is mainly ochre and brown warm colors, blue and green interspersed among them.

(Text by: Liu Yuanfeng)

图：刘元风 Picture by: Liu Yuanfeng

The pattern on offering Bodhisattva's uttariya

on the lower area of the north wall in the main chamber of cave 401 dated to the early Tang Dynasty at Mogao Grottoes

供养菩萨披帛图案

莫高窟初唐第401窟主室北壁下

菩萨的上衣底色为红色，以白色菱格、白色联珠纹和蓝绿对角纹为显花。以白色羽状纹组成主体菱格骨架，主纹为中心环联珠纹，其上下分别装饰绿蓝对角纹。纹样妍丽热烈，花地分明，满地排列，装饰性强，色彩对比强烈，极具视觉冲击力。

联珠纹源于中亚地区的波斯萨珊王朝时期，六世纪中期在中国出现，至唐时期最为盛行，中国西北出土实物中多见此纹样。联珠纹是由大小基本相同的圆形几何点连接排列，形成更大的几何形骨架，然后在这些骨架中填以动物、花卉等各种纹样，有时也穿插在具体的纹样中作装饰带。联珠纹规则有序，富于变化，灵活搭配，自由组合，使其不断发展，样式层出不穷，成为隋唐时期最流行的图案之一。

（文：苏芮）

The background color of Bodhisattva's upper clothes is red, with white rhombic pattern, white beads pattern and blue-green diagonal pattern. The main rhombic frame is composed by white feather pattern; the main pattern is the central ring beads pattern, and the top and bottom are decorated with green and blue diagonal pattern respectively. The patterns are gorgeous and warm, having clear contrast with the background and cover the whole area, very decorative. The color contrast is strong, and the visual impact is very impressive.

Sassanian roundels pattern originated from the Sassanian Dynasty of Persia in Central Asia. It appeared in China in the middle of the 6th century, and became popular in the Tang Dynasty. This pattern is often found in objects unearthed in Northwest China. Sassanian roundels pattern is composed by small roundels to form a larger geometric frame, and then these frames are filled with animals, flowers and other patterns, sometimes interspersed in specific patterns as decorative bands. Sassanian roundels pattern is regular and order, full of changes, flexible collocation and free combination, which makes it develop continuously and keep having new styles endlessly. It had become one of the most popular patterns in the Sui and Tang Dynasties.

(Text by: Su Rui)

图：苏芮 Picture by: Su Rui

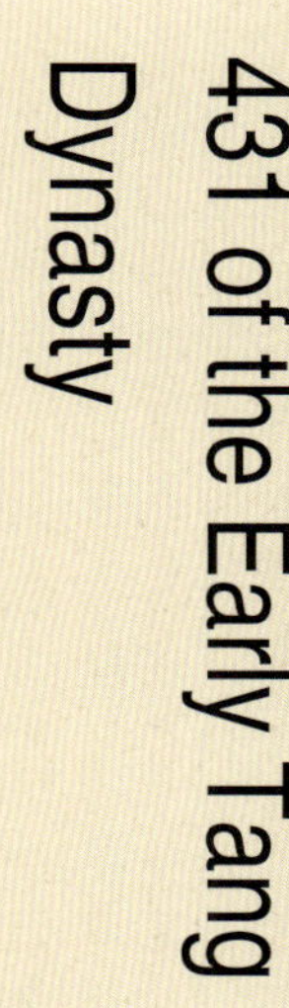

敦煌莫高窟初唐第431窟

Dunhuang Mogao Grottoes Cave 431 of the Early Tang Dynasty

第431窟是始建于北魏时期的中心塔柱式洞窟，后经初唐和宋代重修。主室前部的人字披图案和后部的平棋图案、中心塔柱四面龛内外装饰以及四壁上部的天宫伎乐、中部的千佛和千佛中央的说法图基本为北魏原作。初唐壁画主要位于洞窟的下部，包括主室中心塔柱南、西、北向龛下的卷草纹边饰和释迦多宝佛或说法图壁画，主室南北两壁下部的九品往生十幅、女供养人像和观无量寿经变、男供养人像，主室西壁下部的十六观及供养牛车、马匹、马夫及男女供养人像，东壁门南侧两侧的天王像二身。此窟前室和甬道存宋代壁画，窟前有宋代木构窟檐一座，窟檐前梁存北宋太平兴国五年（980年）题记。

此窟主室南、北、西壁下部绘有初唐供养人多身，按照南壁女供养人、北壁男供养人的方式进行排列。虽然有的壁画人物肤色已经变色，但是服饰穿搭和色彩配置仍然较清晰，从画像数量、艺术水平、身份多样性等方面都展示了世族大家礼佛供养的宏大场面。北壁下部未生怨、西壁下部十六观中的世俗人物服饰，东壁下部天王服饰、中心塔柱下部西、南、北三个向面的佛菩萨服饰细节绘制较为完整丰富，都展现了典型的初唐风尚，为研究初唐时期服饰文化提供了较为真实可靠的图像资料。

（文：崔岩）

Cave 431 is a central pillar type cave built in the Northern Wei Dynasty, which was repaired in the early Tang Dynasty and Song Dynasty. The painting on the gabled ceiling in the front of the main chamber and the painting on the flat ceiling in the back, the interior and exterior decoration of the four niches of the central pillar, the musicians in heavenly palace on the upper area of the four walls, the Thousand Buddhas in the middle area and the preaching scenes in the center of the Thousand Buddhas are basically the original work of the Northern Wei Dynasty. The murals of the early Tang Dynasty are mainly located in the lower part of the cave, including the scrolling grass pattern edge ornaments under the south, west and north niches of the central pillar in the main chamber and the murals of Sakyamuni Buddha and Prabhutaratna Buddha or preaching scenes. The lower part of the north and south walls of the main chamber have ten pictures of nine-fold future life, male and female donors images and Amitayurdhyana Sutra tableau. The lower part of the west wall in the main chamber has Sixteen Observations of Amitabha, ox cart, horse, coachman and male and female donors images. On both sides of the east gate are two bodies heavenly kings. The anti-chamber and the corridor have the Song Dynasty murals. Outside the cave has a wooden facade, there is an inscription on the beam, the fifth year of TaipingXingguo, the Northern Song Dynasty, which dates back to 980AD.

In the lower part of the south, north and west wall of the main chamber, there are many donors dating back to the early Tang Dynasty, arranged in the way of female donors on the south wall and male donors on the north wall. Although some characters' skin color have changed, the clothes and color configuration are still clear. The number of portraits, artistic level, identity diversity and other aspects all show the grand scene of powerful family's Buddhist worship. There is Ajatasattu's story (unborn resentment) on the lower part of the north wall, the clothes of the secular figures in the Sixteen Observations of Amitabha on the lower part of the west wall, the clothes of the heavenly king on the lower part of the east wall, and the clothes of Buddhas and Bodhisattvas on the west, south and north lower part of the central pillar; all the drawing of the clothes' details are relatively complete and rich, which shows the typical fashion of the early Tang Dynasty, and provides a relatively real and reliable image information for the study of clothes culture in the early Tang Dynasty.

(Text by: Cui Yan)

The female donor's clothes on the lower area of the south wall in the main chamber of cave 431 dated to the early Tang Dynasty at Mogao Grottoes

女供养人像服饰 莫高窟初唐第431窟主室南壁下

第431窟建于北魏晚期，初唐将窟底下挖重建，补充了部分壁画。主室南壁在初唐补画的是十幅九品往生、一身比丘尼、二十二身女供养人、四身侍从。

二十二身女供养人的服饰虽有色彩差异，但造型风格统一，可窥见初唐女装风貌。在绘画整理时，选取了其中最有特点的一个形象——壁画中的这位供养人肤色白皙，脸形略丰腴，头梳螺髻，发型梳理得紧实有型，为初唐的典型风格。上身穿深色窄袖上襦，肩搭披帛，一端塞入长裙，另一端垂至膝下。下着高腰间色长裙，裙长及地，裙色红绿相间，壁画中腰节部位为白绿相间，或许是红色颜料脱落所致。壁画上的鞋履形态已斑驳，参照同窟供养人形象，并依照残形推断应为花头履。

间色条纹裙始于魏晋、兴于唐，至盛唐以后逐渐被色彩艳丽的花纹裙取代。早期间色的条纹较宽，之后呈渐窄的趋势。这身供养人穿着的间色裙条纹极窄，在初唐尚不常见。

纵观二十二身供养人的服饰，整体廓型修长，体现了初唐“襟袖狭小”的特点，从洞窟现存壁画的人物整体形态来看，几乎没有佩戴任何簪环首饰，总体风格极为清减，这与之后流行的丰硕饱满的服装造型、繁缛夸张的首饰配件形成了鲜明对比。

（文：李迎军）

Cave 431 was built in the late Northern Wei Dynasty. In the early Tang Dynasty, the bottom of the cave was excavated downward and painted some new murals. On the south wall of the main chamber has ten pictures of nine-fold future life, one bhiksuni, twenty-two female donors and four attendants which all dated back to the early Tang Dynasty.

Although the 22 female donors' clothes have color differences, their style is same, which can reveal the style of woman's clothes in the early Tang Dynasty. In the process of effecting drawing, one of the most characteristic images is selected. This woman has white complexion, a slightly plump face, a snail shell bun and a tight hairstyle, which was a typical style of the early Tang Dynasty. She wears a dark narrow sleeves upper coat and a uttariya over her shoulders. She tucked one end into the long antariya and the other end down to her knees. A long striped antariya with high waist is on the lower body, and drape on the ground. The colors of the antariya are red and green, the waist part is white and green. Maybe the red pigment fell off so it looks white now. The shoes on the mural are mottled, referring to the image of other donors in the same cave and the remnant, we presume that the shoes should be decorated with flower heads.

The striped antariya appeared in the Wei and Jin Dynasty and was gradually replaced in the high Tang Dynasty by colorful flower pattern antariya. In the early stage, the stripes were wider, and then it became narrower. This donor wears striped antariya, the stripes are extremely narrow, not common in the early Tang Dynasty.

Looking at the 22 donors' clothes, the overall size is slender, which reflects the characteristic of "narrow lapels and sleeves" in the early Tang Dynasty. The figures in this cave almost have no jewelry, and the overall style is extremely simple, which is a strong contrast with the rich and full clothes styles and exaggerated jewelry accessories in later times.

(Text by: Li Yingjun)

图：李迎军 Picture by: Li Yingjun

The male donor's clothes on the lower area of the north wall in the main chamber of cave 431 dated to the early Tang Dynasty at Mogao Grottoes

男供养人像服饰
莫高窟初唐第431窟主室北壁下

第431窟的北壁上，在北魏原有的禅定佛、千佛、天宫伎乐之下，初唐又绘制了未生怨故事，以及四身供养比丘、二十六身男供养人、八身侍从。

这二十六身供养人姿态相似，皆双手拱于胸前，神情肃然，态度虔诚。他们的服装也几乎一致——每人都头裹软脚幞头，穿圆领窄袖襕衫，腰束革带，足蹬乌皮靴。

隋唐时期皇帝与官员的服装大约分为两类：一类是传统的汉式冠冕衣裳，主要是用作冕服、朝服等礼服和较为简化的公服；另一类是常服，主要是圆领袍和幞头。按《新唐书》所载："至唐高祖，以赭黄袍、巾带为常服。"后来礼服的使用范围逐渐缩小，而重实用功能的常服进一步扩大化，实际上已取代了朝服和公服的地位。如《旧唐书》所说："自贞观以后，非元日冬至受朝及大祭祀，皆常服而已。"圆领袍的下端有时会拼接一块横襕，象征古代上衣下裳的分制，称为"襕衫"或"襕袍"。第431窟壁画中的男供养人皆穿圆领窄袖襕衫，尽管衣身的着色已大片脱落，且色彩氧化变色严重，但衣摆上横襕的结构线仍然清晰可见。

（文：李迎军）

On the north wall of cave 431, below the meditation Buddha, Thousand Buddha motif, and heavenly palace musicians dating back to the Northern Wei Dynasty, people in the early Tang Dynasty drew stories of Unborn Resentment (Ajatasattu story), as well as four offering bhikkhus, twenty-six male donors, and eight servants on the lower area.

These 26 donors were similar in posture; their hands hold in front of their chests, with a solemn look and a devout attitude. Their clothes are almost the same — everyone wears soft feet Futou, round collar narrow sleeve robe, leather belt around waist, and black leather boots.

During the Sui and Tang Dynasties, the clothes of emperors and officials can be divided into two categories: one is the traditional Han style crown clothes, which are mainly used for ritual clothes, court clothes and some formal dress; the other is the regular clothes, which are mainly round collar robe and Futou. According to *the New Book of the Tang Dynasty*: "In time of the emperor Gaozu of the Tang Dynasty, he often wore ochre yellow robe and bands." Later, the usage of ritual dress gradually narrowed, and the regular dress emphasized on practical function and became more popular; in fact, it has replaced the status of court dress and official dress. As *the Old Book of the Tang Dynasty* wrote: "since the year of Zhenguan, except the first day of winter solstice and big sacrifice, people all wear regular dress." The lower end of the round collar robe is sometimes stitched with a piece of horizontal cloth（横襕）, which symbolizes the ancient Yi and Shang（上衣下裳）, and is called "Lanshan（襕衫）" or "Lanpao（襕袍）". In cave 431, all the male donors wear round collar narrow sleeves Lanshan. Although the colors of the clothes have fallen off and seriously oxidized and discolored, the structural lines of the horizontal cloth are still clearly visible.

(Text by: Li Yingjun)

图：李迎军 Picture by: Li Yingjun

服装复原：楚艳

文字说明：崔岩、楚艳

摄影：陈大公、卢硕

造型：林颖、高丽环、赵璞

模特：王禹丁、王青年、张巨鹏、高诗雨、刘浩然、周文政、李显辉

Costume Reproduction: Chu Yan

Text: Cui Yan, Chu Yan

Photo: Chen Dagong, Lu Shuo

Make-up: Lin Ying, Gao Lihuan, Zhao Pu

Model: Wang Yuding, Wang Qingnian, Zhang Jupeng, Gao Shiyu, Liu Haoran, Zhou Wenzheng, Li Xianhui

敦煌服饰艺术再现

The Reappearance of Dunhuang Costume

天女服饰的总体款式为襦裙装：内着白色中单，曲领半露；上着交领大袖裙襦，袖边为靛蓝色；下着浅绿色长裙，盖至脚面；外搭朱红色袖缘羽毛半臂，均系于裙腰之内；腰系石绿色长带，长至脚踝，围联珠纹刺绣蔽膝；着“凹”字形翘头履。此款服饰面料采用真丝花罗、丝麻提花、丝棉提花等面料，暗提花图案来自云纹及花草等唐代纹样。刺绣图案主要集中于蔽膝和羽饰半臂上，包括联珠团花纹、四瓣花纹、卷草纹等。服饰色彩以白色、石绿色、靛蓝色、朱红色为主，清新雅丽。天女发髻高耸，佩戴镶嵌珠翠的对凤纹和如意云头金钗，飘扬的鬓发和轻执的麈尾更加突出了天女仪态万千、闲庭信步的自在风度。

The overall style of the celestial woman's dress is Ru and long skirt: she has white Zhongdan inside, curved collar half exposed, covered by large sleeve skirt Ru with cross collar and indigo blue on the sleeve trimmings; the lower body wears light green long skirt, covering to feet; on the upper body wears a red half sleeves shirt with feather shaped trimming, all tied in the waistband; the waistband is malachite green, reaches to ankles, the legs front is covered by Bixi with beads pattern embroidery;the shoes are "concave" shaped. This dress materials are made of silk yarn, silk hemp jacquard, silk cotton jacquard and other fabrics. The hidden jacquard pattern comes from the patterns of cloud and flowers and grass in the Tang Dynasty. Embroidery pattern mainly focuses on Bixi and feather shaped half sleeves shirt, including beads pattern, four petals pattern, scrolled grass pattern, etc. The dress colors are mainly white, malachite green, indigo blue and vermilion, which are fresh and elegant. The bun of the heavenly woman is high, the hairpin is inlaid with green stones and has double phoenix pattern and designed as Ruyi style and cloud shaped ends. The fluttering hair and the fan holding in her hand highlight the heavenly woman's graceful demeanor and peaceful inner world.

The Female Donor's Costume Reappearance

from Cave 329 dated to the early Tang Dynasty in Dunhuang Mogao Grottoes

女供养人服饰艺术再现

敦煌莫高窟初唐第329窟

这身女供养人服饰总体紧窄贴身，时尚简约。上着紧身圆领窄袖小衫，外套半臂；下着高束腰间色长裙。本套服饰面料采用真丝素缎、真丝花罗、丝棉提花等面料制作，间裙饰印花图案。服饰色彩以浅黄色、褐色、朱磦等暖色调为主。头饰及项链由仿珊瑚、玛瑙、绿松石等天然石珠串制而成。手持莲花安然跪坐的姿态体现其虔诚供养的神情。

This female donor's dress is tight and suits to the body, which is fashionable and simple. The upper body wears tight round neck narrow sleeves shirt, coated by half sleeves shirt; the lower part wearing a striped long skirt with high waist. This set of dress fabrics are made of silk plain satin, silk jacquard, silk cotton jacquard, etc., and the skirt is decorated with printing pattern. The colors of the dress are mainly light yellow, brown, light red and other warm colors. The headdress and necklace are made of fake coral, agate, turquoise and other natural stone beads. The kneeling posture with lotus in hand reflects her pious offering.

听法王子服饰艺术再现

敦煌莫高窟初唐第332窟

The Prince's Costume Reappearance

from Cave 332 dated to the early Tang Dynasty in Dunhuang Mogao Grottoes

图中双手合十的听法王子服饰具有鲜明的域外色彩。头戴尖顶织锦帽配刺绣饰边，着窄袖长袍，足蹬尖头乌皮靴。根据初唐图案的特点，重新整理并以刺绣工艺呈现了装饰于长袍缘边的半团花纹。服饰采用丝麻菱格纹提花面料，大面积色彩为棕褐色，点缀上繁密绚丽的刺绣缘边图案，显得华贵敦厚、沉稳雅致。

In the picture, the clothes of the prince with his hands folded have distinctive foreign style. He wears a tapestry hat with embroidered trim, a long gown with narrow sleeves and pointed black leather boots. According to the characteristics of the patterns in the early Tang Dynasty, the semi medallion flower pattern decorated on the edge of the robe are rearranged and presented by embroidery technique. The dress is made of silk and hemp jacquard fabric with rhomboid pattern. The main color is brown, decorated with rich and gorgeous embroidery edge patterns, which makes it look luxurious and elegant.

听法王子服饰艺术再现

The Prince's Costume Reappearance

敦煌莫高窟初唐第332窟

from Cave 332 dated to the early Tang Dynasty in Dunhuang Mogao Grottoes

此身王子的服饰承袭了上衣下裳的传统。头戴莲花状小冠，上身内穿曲领中单，外套直领大袖衫，腰间以阔丝带系结，带端垂于前，长至膝下，裙前垂蔽膝，脚穿高齿履。服饰采用几何纹样暗提花丝织面料，色彩以浅石绿色、浅蓝色、米白色、暖灰色为主，显得低调内敛，儒雅有度。

The prince's dress inherits the tradition of upper Yi and lower Shang. He wears a small lotus shaped crown on his head, a curved collar Zhongdan on his upper body, a large sleeve shirt with a straight collar as his coat, a broad ribbon tied around his waist, with the ends hanging to the front and reaching below the knees. The front of the Shang is covered by Bixi, with high teeth shoes on his feet. The dress is made of hidden jacquard silk fabric with geometric patterns, and the colors are mainly light malachite green, light blue, rice white and warm gray, showing low profile, introverted and elegant.

The Disciple's Costume Reappearance
from Cave 333 dated to the early Tang Dynasty in Dunhuang Mogao Grottoes

弟子服饰艺术再现
敦煌莫高窟初唐第333窟

右图佛陀身边的听法弟子身着常规的比丘服饰，威仪端庄。内着僧祇支，外披田相纹袈裟，下着多褶裙，足蹬红色高齿履。服饰采用暗提花丝织面料，印花图案较为丰富，主要包括袈裟上的五瓣花纹和裙身上的四瓣散花纹，尤其是袈裟上的花纹造型形似梅花，以黑色、石绿色和红色等单色轮廓式平涂法表现，具有平面化的装饰意味。

In the picture, the disciple of Dharma listening beside the Buddha is dressed in traditional bhikkhu costumes, dignified and solemn. Inside is a monk's Sankaksika, outside is a kasaya with field pattern, under is a pleated dhoti, and a red high teeth shoes on his feet. The dress is made of hidden jacquard silk fabric with rich printing patterns, mainly including five petals flower on the kasaya and four petals flower on the skirt. In particular, the pattern on the kasaya is shaped like plum blossom, and is expressed by monochrome contour flat painting method, with black, malachite green and red to depict, has a complanation decoration taste.

The Male Donor's Costume Reappearance

from Cave 375 dated to the early Tang Dynasty in Dunhuang Mogao Grottoes

男供养人服饰艺术再现

敦煌莫高窟初唐第375窟

此身供养人所着为初唐时期男子的典型装束。头裹黑色软脚幞头，穿绯色圆领袍服，腰束紫带，足蹬乌靴。服饰采用团花纹丝毛暗提花面料，主体色彩为绯红，反映了唐代官服阶位的史实。手持莲花的挺拔身姿显得恭谨干练，表现了虔诚供养的神态。

This donor is wearing the typical costume of a man in the early Tang Dynasty. His head is covered by black soft feet Futou, and dressed in scarlet round neck robes, with purple belts around his waist and black boots on his feet. The clothes material is hidden jacquard silk fabrics with medallion flower pattern; the main color of the dress is crimson, which reflects the historical facts of the rank of official dress in the Tang Dynasty. The holding lotus flower and upright posture appears to be respectful and capable, showing the expression of devcut offering.

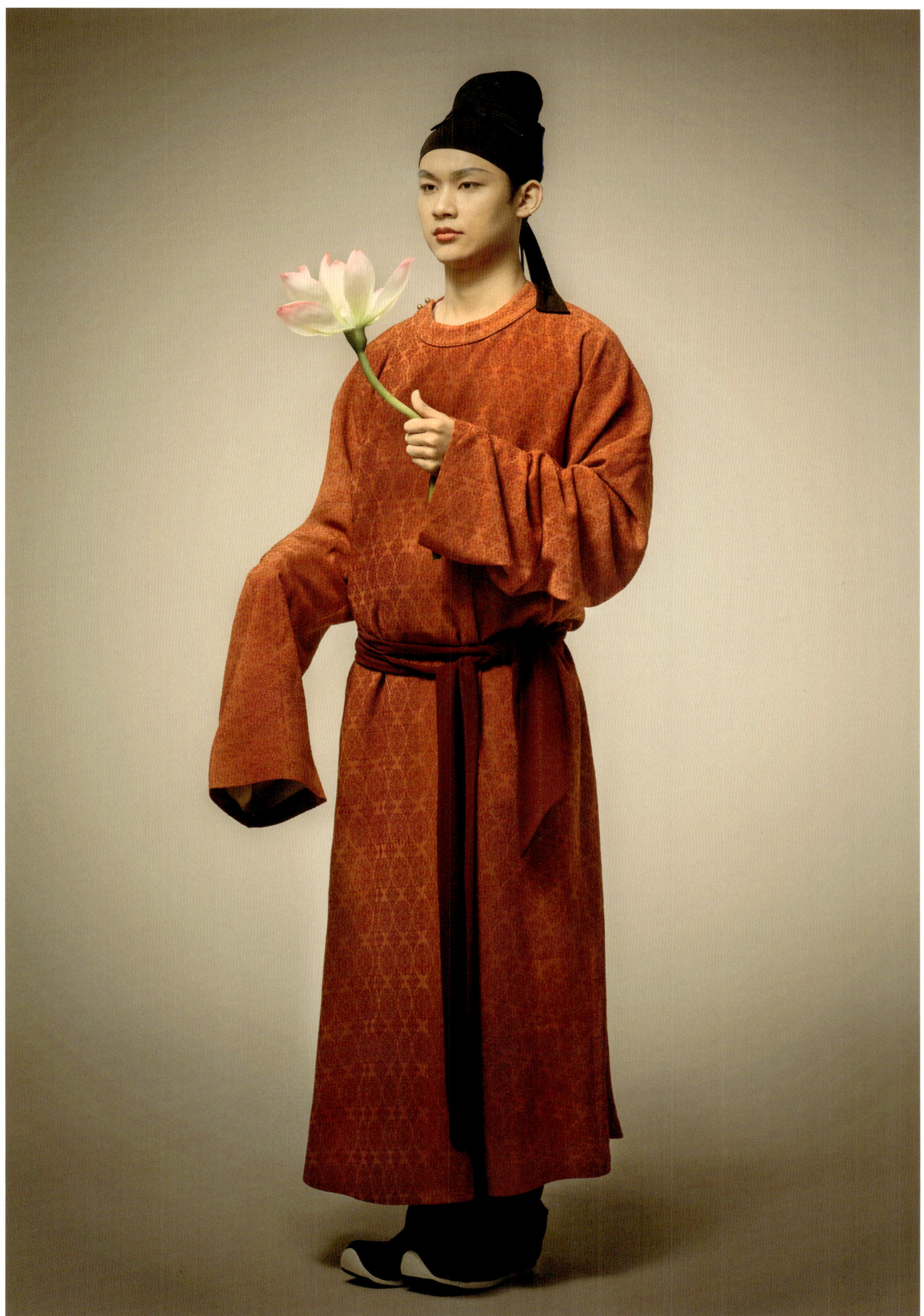

女供养人服饰艺术再现

敦煌莫高窟初唐第375窟

The Female Donor's Costume Reappearance

from Cave 375 dated to the early Tang Dynasty in Dunhuang Mogao Grottoes

画面中身材颀长的女供养人身着初唐时期典型的女子服饰。头梳高髻，上着窄袖交领襦，披帛自双肩绕臂自然垂于两侧；束腰带，下着高腰间色长裙，着绣花履。服饰色彩讲究淡雅明丽，以浅石绿和石青色为主，搭配以茜色的披帛和腰带，裙腰处刺绣卷草纹，在对比变化中追求和谐统一的效果。

The tall female donor in the picture is dressed in the typical female dress of the early Tang Dynasty. The hair is combed in a high bun; the upper body wears a narrow sleeve clothes with cross collar. The sash is draped from the shoulders around the arms and naturally hangs on both sides; a belt on waist, a long skirt with high waist on the lower body and embroidered shoes on feet. The color of the dress is elegant and bright, with light malachite green and azurite as the main color. It is matched with sash and belt in alizarin color. The waist of the skirt is embroidered with scrolling grass pattern. It pursues the effect of harmony and unity by the contrast and change.